Meeting the Stress Challenge

A training and staff development manual for social welfare managers, trainers and practitioners

Neil Thompson, Michael Murphy and Steve Stradling

with Paul O'Neill

MEETING THE STRESS CHALLENGE

Russell House Publishing Limited

First published as a wirobound A4 manual in 1996, and as an A5 paperback in 1998 by:
Russell House Publishing Limited
4 St. George's House
Uplyme Road
Lyme Regis
Dorset DT7 3LS

British Library Cataloguing-in-Publication Data:
A catalogue record for this manual is available from the British Library.

ISBN: 1-898924-47-3

Text design and layout by: Jeremy Spencer, London
Printed by Hobbs the Printers, Totton

Contents

Welcome...

...to *Meeting the Stress Challenge*, a guide for managers, trainers and practitioners at all levels within social welfare. Its purpose is twofold.

- First, it provides an informative overview of what stress is, what causes it, why it is important and what can be done about it. It is therefore an important resource for you to draw upon in facing up to the challenges presented by work-based stress.
- Second, the manual can be used as the basis of providing briefing and training for all staff, to help you in the collective battle to meet the stress challenge.

The fact that you have already begun to invest time in reading the manual indicates a commitment to meet this challenge. The manual therefore contains suggestions for training exercises you can do with a group of staff as part of a training course or staff development programme, together with exercises that can be done on an individual basis.

The intended readership of the manual is social welfare staff, broadly defined. This includes social workers, probation officers, social care workers and youth and community workers. And, within those professional groups, it includes practitioners, managers and training staff.

The manual will not provide you with all the answers or solve all your problems (if only life were that simple!) but it should give you a firm foundation on which to build, and should equip you to make major steps forward in dealing with the problems of stress. Although the manual deals primarily with issues of work-based stress, stress, unhappily, does not recognise the boundaries between 'home' and 'work' or 'private' and 'public' and therefore we include consideration of personal as well as professional stress. As we shall emphasise towards the end of the manual, you owe it to yourself and to your staff to make a positive contribution to meeting the stress challenge. So, before you launch into the manual, all that remains is for us to wish you well in your efforts. Good luck!

About the authors

Neil Thompson teaches in the School of Health and Community Studies at North East Wales Institute. He has worked as a social worker, team leader and training officer. He is the author of a number of social welfare books, including *Theory and Practice in Health and Social Welfare* (Open University Press, 1995) and *People Skills* (Macmillan, 1996).

Michael Murphy is a multidisciplinary child protection trainer and the coordinator of a staff care scheme. He has worked as a social worker and a social work tutor. He is the author of Working Together in Child Protection (Arena, 1995) and *The Child Protection Unit: Its History, Function and Effectiveness in the Organization of Child Protection Work* (Avebury, 1996).

Steve Stradling is senior lecturer in psychology at the University of Manchester. He has a broad experience of teaching, research and consultancy in the field of psychology. He is the co-author (with M.J. Scott) of *Counselling for Post-traumatic Stress Disorder* (Sage, 1991).

The three main authors of this manual are also the authors of *Dealing with Stress* (Macmillan, 1994).

Paul O'Neill is a social worker in Northern Ireland. He has an MSc in Applied Psychology on the subject of stress in social work.

The authors are able to provide training and consultancy services in relation to stress management and staff care issues. For further information, contact:

Ashley Maynard Associates
1 Worcester Road
Bangor on Dee
Wrexham LL13 0JB
Tel: 01978 780296
Fax: 01978 781117
Email: neil@maynard.u-net.com
http://www.mjordan.demon.co.uk

Preface

This manual stems from earlier work undertaken by the Stress In Social Work Research Group, of which the current authors have been members since its inception in 1988. From the various research and consultancy projects we conducted, certain recurring themes began to develop, important themes that we felt should be brought to the attention of a wider audience. These key themes are:

- **The damage that stress can do.** The overall costs of stress are of immense proportions but rarely fully appreciated, even by those who are heavily affected by them.
- **Our responsibilities.** We all have a responsibility to play a part in the collective struggle to tackle stress and develop a positive and supportive work environment, with managers and trainers having broader organisational responsibilities as well as their own individual ones.
- **The central role of training and staff development.** An awareness of stress issues and the steps necessary to tackle them can very effectively be developed through training events and a wider focus on staff development (for example, supervision).

The recognition of these three key themes led us to focus our attention on the training and staff development aspects of dealing with stress, and this manual is the product of our decision to focus our attentions in this way.

In emphasising these aspects of the stress challenge we are most certainly not suggesting that these are the only sets of issues that need to be addressed. However, we do see them as a crucial part of developing a strategic approach to stress management and related issues.

Using this manual – publisher's note

The initial publication of *Meeting the Stress Challenge*, as an A4 wiro-bound manual, proved to be a very useful and successful resource for all sorts of organisations and agencies. To enable this valuable material to be made available to a wider audience, we have decided to produce this smaller, and more economical, A5 edition.

Meeting the Stress Challenge is designed to be used in several ways. As the authors explain in the introduction, it can help identify and alleviate stress amongst both managers and practitioners. The introduction also clarifies the different, but closely connected, purposes of each part; we have added this note to outline how best you can use the material in this manual.

Part One: Understanding Stress is for people at all levels in social welfare to read and ponder. While so doing, the sub-sub sections entitled, 'Myths About Stress' and '25 Ways to Stress out Your Staff' allow you to make notes and compile thoughts.

Part Two: Stress Management Training is useful for any group who want to work together to alleviate stress. For example, it will help trainers and managers who are conducting staff development activities. You will find that the overhead transparency templates at the end of the manual will help in this work. By using the zoom or enlarge feature, the forms and OHP templates can be photocopied and then used at their original A4 size.

Part Three: Recognising Stress, like Part One, is for reading and reflection.

Part Four: Tackling Pressure and Stress is, in the author's words, "meant for the individual member of staff and can be used as the basis of a handout or explanatory leaflet for whole team training or for specific stress workshops, training courses or other staff development activities".

The **Conclusions** set out on separate pages the lessons for managers, trainers and practitioners, and are useful supplements or reinforcements for your work together.

Introduction

Stress has been recognised as a serious problem of working life, as the growing literature on the subject confirms (see the *Finding Out More* section below). That stress causes harm to individual staff, teams and whole social welfare organisations is an important fact to acknowledge if its negative effects and destructive potential are to be prevented, kept to a minimum or responded to effectively and helpfully. This, then, is the starting point – the recognition that stress is a problem both for the individual and for the organisations they operate in. It is no good taking a 'be tough' attitude for, as we shall see below, this can actually make matters worse. Trying to sweep the problem under the carpet is not only unhelpful but can, in some circumstances, amount to 'lighting the blue touch paper'. That is, at best it delays the problem but, at worst, can lead to a very damaging situation for all concerned.

Bethan was a young, well-thought-of manager in the Coverdale department. She had been asked to go in to help manage the south west district team, whose manager was absent through stress. On the Monday morning she called an emergency team meeting. The atmosphere was fraught — the team members seemed agitated, one was openly crying. Kindly, but firmly, Bethan told the team that there was no alternative but to push on together to get the job done. By Thursday another two members of staff had gone off on sick leave. Even worse, Bethan began to feel a rising personal anxiety that she would not be able to cope. On Friday morning Bethan called another emergency team meeting: 'We need to start again. Could you start by telling me how it is, why it is so bad and what I can do to help'.

An important message to emerge from this manual, then, is the need to understand:

- what factors lead to stress
- what helps in dealing with it, and
- what does not.

Indeed, these can be seen to be closely linked to the main aims of the manual:

1. To provide an overview of the key issues relating to stress – so that all staff can work from an informed basis;

2. To offer guidance on how managers can help staff to deal with stress positively and constructively – so that they can play an active part in supporting the staff under their command;
3. To assist staff responsible for training and staff development to organise and run activities geared towards developing stress management knowledge and skills;
4. To provide guidance on what each individual staff member can do to play a positive part in dealing with stress;
5. To explore the strategies that can be used to prevent stress; to deal with it when it starts to get out of hand; and to handle the consequences or aftermath of stress.

In order to meet these aims, the manual is arranged in five sections as follows:

In Part 1, we look at *understanding stress.*

Why is stress important?
Here we see the harmful effects of stress and the ways in which it can cause major problems for all concerned. Our intention here is to help you develop a clearer picture of the costs that are associated with stress, not only financial costs but also personal, organisational and social costs.

The importance of stress in social welfare
While stress can be seen as a very problematic phenomenon in any organisation, there are particular features of social welfare work that make stress a particularly significant issue. This section therefore addresses the particular pressures that social welfare workers face.

What is stress?
It is important to define exactly what we mean by stress, consider what causes it, and challenge some of the myths surrounding it. The term 'stress' is used in different ways by different people and is often used loosely and uncritically. This section seeks to clarify precisely what we mean by stress.

Meeting the challenge
Stress presents a major challenge for workers, managers and trainers but, for the latter two groups, the challenge is a double one – involving dealing with *one's own* pressures and supporting one's staff in coping with *their* pressures. In this part of the manual, this challenge is explained and a number of ways of dealing with it are explored. The emphasis is on practical measures which are relatively simple to put into practice but which can be very effective in reducing stress and increasing job satisfaction.

In Part 2, a number of suggestions are made for *exercises* that can form the basis of a training course, or can be used in staff meetings or other staff development activities. The basis of each exercise is explained with a view to enabling trainers or managers to put them into practice as they stand or adapt them to their own requirements.

In Part 3, we address the question of *recognising stress*. This section of the manual details a number of common sources of workplace stress and a number of common signs of individual distress to assist you in trying to work out: 'what exactly is our problem here?'. Accurate assessment of the individual and organisational problems is vital to devising effective, durable and acceptable solutions. This section also illustrates the sources of job satisfaction that have been identified from recent research, since the alleviation of distress and the removal of stressors should not be the end of the process but clears the way for the enhancement of individual competency, enabling staff and managers to deliver a better service to their clientele.

In Part 4, the question of *tackling pressure and stress* is the main focus of discussion. This comprises a set of practical suggestions to help individual staff play their part in dealing with stress – the organisation as a whole has a part to play in meeting the stress challenge but so too have individual members of staff.

The *Conclusion* pulls together the main themes and summarises the main points covered, and, in so doing, emphasises the need to take stress issues very seriously.

In these times of rapid change and increasing pressure on local government, voluntary and independent agencies, it is not surprising that stress is being recognised as an important problem which deserves closer attention. This manual is designed to help you do that, to play a positive role in making the experience of work less fraught and more satisfying for all concerned – with the added bonus of making your own job more positive, constructive and satisfying in the process.

There is, then, much to be gained in seeking to meet the stress challenge and, as we shall see in the next section, a great deal to be lost if the problems of stress are 'swept under the carpet' and not tackled head on. The first step towards meeting the challenge is to understand just why stress is so important, why we cannot afford to ignore it.

Understanding stress

Why is stress important?

The short answer is: 'It hurts and it costs'. That is, stress has a number of significant costs for all concerned:

For staff:

- poor health
- low morale and motivation
- less job satisfaction
- strained relationships
- lower confidence and self-esteem
- irritability
- irrational behaviour
- low mood/mood swings
- feelings of isolation/alienation
- anxiety
- impatience
- increased use of 'soothing' substances
- poor work output
- intrusive tearfulness.

PRACTICE FOCUS 1.1

From the minute Clare came into her supervision session, it was obvious to her supervisor that something wasn't right. In the middle of a discussion on a relatively routine case she burst into tears: 'I just don't know what is going on. I just can't seem to get anything done and I keep bursting into tears all over the place'.

For family and friends:

- strained relationships
- having to deal with low mood/mood swings
- irritability/impatience
- feeling cut off
- poor performance as partner/parent/family member.

PRACTICE FOCUS 1.2

Terry had had the misfortune to be working on a case that had 'gone wrong'. A complaint had been made by the service user and the local MP. The complaints procedure was long, complex and tortuous. Terry was explaining to his supervisor what it felt like at home: 'On the surface everything is the same – I look the same and so do they – but I just can't talk to them or touch them in any way. It's like living in a bell jar'.

For service users, clients or customers:

- a lower level of service (both quality and quantity)
- delays in receiving a service
- lack of confidence in staff
- refusal of access to services
- unhelpfulness/irritability/impatience shown by staff
- an increased likelihood of mistakes being made
- a reluctance to make demands on over-burdened staff.

PRACTICE FOCUS 1.3

Mrs Beckett had brought in some forms that her social worker had asked her to fill in. She was kept waiting a long time at the front desk. In the end a young person took the forms from her without speaking. She looked at them and handed them back immediately: 'Don't you know we don't deal with that here, you need to go across the road. Next!'. Mrs Beckett, who did not know where 'across the road' was, went home, upset, forms still in her possession.

For teams:

- increased sickness, absenteeism and staff turnover
- increased burden on remaining staff
- a culture of tension, frustration and negativism
- reduced output and more 'covering up'.

PRACTICE FOCUS 1.4

Coltown Community Support team was a project team offering a service to people with learning difficulties living in the community. The team's deputy and administrator were off on long-term sick leave. The team manager was 'acting up' elsewhere in the agency. The team gradually became very inward-looking, the culture cynical and negative. Discussions within the team room were full of phrases like: 'Well, they would say that, wouldn't they!'; 'It just seems like the thin end of the wedge' and 'It's the same old story every time'.

For organisations:

- higher sickness rates and absenteeism
- lower productivity
- possible industrial relations problems (grievances, disciplinary matters and so on)
- higher rates of staff turnover
- greater obstacles to effective service provision.

PRACTICE FOCUS 1.5

Coverdale Youth Service was undergoing its annual round of 'cuts'. At the same time, a major dispute had broken out between the senior staff and their director. In Coverdale, grievances multiplied, over one third of the staff were off on sick leave. In the centres, the atmosphere was openly depressed; very little youth work and no developmental work was being undertaken.

It has been estimated that, in the USA occupational stress costs the nation about ten per cent of GNP per annum (Cooper, 1995a). The Confederation of British Industry (CBI) has calculated the cost to the UK economy of stress-related sickness absence to be in the order of £5.3 billion (Fingret, 1994). These costs make stress a major economic problem and underline the need to take it seriously – the need to move away from the common tendency to pretend that stress does not exist or is not a problem people should own up to.

A further reason why stress is important is that it is self-perpetuating. Once a person is weakened by stress, his or her ability to cope with other pressures tends to be reduced, and this can lead to a vicious circle developing – a 'cycle of stress' (Thompson, 1991). Similarly, stress can be 'infectious'. If one person in a staff team experiences undue stress, this in turn can place an increased burden on other people in that staff group, for example if work tasks have to be reallocated to colleagues. If these extra tasks then produce a further set of pressures for staff, the problem can become compounded.

A team's five staff were coping reasonably well with their workload and were not experiencing major problems. However, one day, Peter, one of the staff was threatened with violence while carrying out his duties. The incident was not taken seriously by his managers and they tried to laugh it off as 'just one of those things'. Peter, however, felt vulnerable, unsupported and undervalued. A few days later, a medical certificate was submitted to confirm that he was to be on sick leave for two weeks as a result of 'nervous exhaustion'. During that time the remaining staff had an average additional 25 per cent workload in order to keep the service running. This placed them under a great deal of strain, but they managed to cope. However, at the end of the fortnight, a further certificate was received from Peter, this time for four weeks. The staff team began to feel despondent and wondered whether Peter would ever return. Consequently, Pam, who had been experiencing marital difficulties recently, reached the point where she could not cope any further. She too went on sick leave, thereby leaving three staff to cope with the work of five – a total workload increase of 40 per cent. The three remaining staff now felt under immense pressure and began to doubt whether they would ever see light at the end of the tunnel. The whole situation had become very fraught and dangerously close to breaking down altogether. (Thompson, 1996, p. 31)

Practice Focus 1.6 raises a number of important points about stress and why it is an important problem to tackle:

- 'critical incidents' (threats of violence, for example) can play a major part in producing stress;
- an inappropriate management response can be significant in exacerbating stress;

- stress 'begets' stress – problems can multiply; stress often manifests itself as sickness absence – either directly as 'nervous exhaustion' or indirectly in a variety of ways: 'flu, stomach upset, migraine;
- stress at work often combines with stress at home – the 'home-work interface'.

These are all important issues – and ones that managers need to be aware of if they are to play a positive role in meeting the stress challenge. It is therefore vitally important that managers have a good understanding of stress so that its destructive effects can be guarded against.

The importance of stress in social welfare

Social welfare work is an intrinsically pressurised form of activity and, of course, such pressures can easily overspill into stress. At the same time, there are a number of reasons why stress can have a particularly negative impact on the delivery of social welfare services.

■ The pressures of social welfare work

At its most simple, social welfare has a particularly strong potential for stress because it is primarily concerned with people rather than with goods, wealth or knowledge. This 'people focus' involves several intrinsic factors that can lead to stress:

Uncertainty

People, families and communities behave in ways that are unpredictable. Social welfare workers can never achieve complete predictability and must therefore be prepared to work in reactive ways in response to unplanned events and crises. In recent years this uncertainty has been increased by forced crises of uncertainty involving reorganisations, 'downsizing', service cuts and redundancies. Thus, the characteristic, day-to-day uncertainty of social welfare work has been multiplied by the uncertainty of forced change brought about by fiscal and political (rather than professional) pressures.

Barrowdale Social Services Department had been in the process of reorganising for almost six years. Two directors had come and gone; the council had changed control and a range of different plans and strategies had been considered. To the staff, it seemed that the 'why' and the 'where' of the reorganisation had been forgotten a long time ago – only to be replaced by the desperate need to reduce expenditure. It was clear that they were suffering from a form of 'reorganisation fatigue'.

Vulnerability, pain and suffering

The people that social welfare workers seek to work with tend to be those who, in societal terms, could be deemed to be the most vulnerable and the most disadvantaged. Furthermore, social welfare will have its greatest involvement at times of particular pain and suffering for these clients. One of the most basic sources of stress in social welfare is in the mirroring of this vulnerability and suffering.

PRACTICE FOCUS 1.8

Jasmin had just accompanied Theresa, a young woman, to a planning meeting to discuss possible future options. At that meeting, however, Theresa had been strongly castigated and rejected by her family – a very fraught and distressing situation for all concerned. Later that night, much to her partner's concern, Jasmin was discovered on her own in their spare room in a very distressed state. Theresa's rejection had struck a very deep, negative chord with Jasmin, a clear example of the emotional demands on her.

Poverty

Poverty is a condition which is usually present in clients of social welfare. Poverty cuts down possible solutions to social problems and increases clients' and workers' feelings of helplessness. Although social welfare workers are likely to be accustomed to poverty through their day-to-day exposure to its costs and consequences, the last quarter of the twentieth century has arguably seen a substantial increase in the extent and quality of that poverty, added to a decrease in optimism that it can be overcome (see, for example, NCH, 1993). Poverty has long been recognised as grinding; now it is frequently perceived as permanent and hopeless.

Discrimination and oppression

Social welfare often involves working with people who have experienced or continue to experience discrimination and oppression (Thompson, 1993). This brings three sets of pressures for staff:

1. Coming face to face with the destructive effects of racism, sexism, ageism and so on can, in itself, bring additional pressures to the role – the sheer weight of oppression in clients' lives can be overwhelming at times.
2. For staff who have themselves experienced discrimination there are additional pressures that can arise, not least the emotional pressures of painful memories of one's own being triggered by encountering similar situations that clients are struggling with.
3. The need to develop forms of practice that do not reinforce discrimination and oppression and actually challenge inequality and injustice is one that is increasingly being recognised as a basic requirement of good practice. However, it must also be recognised that such a need brings with it an additional set of pressures for staff.

Care versus control

Society frequently demands that social welfare staff undertake two contradictory tasks with regard to their clients. This is particularly relevant in the area of 'care or control'. On the one hand staff need to care for clients; on the other they need to control client behaviour on behalf of society. This conflict of demand can be particularly stressful, and is an intrinsic feature of work in the areas of child protection, mental health and youth justice (see the discussion below of 'the politics of welfare').

PRACTICE FOCUS 1.9

Frances was a social worker in a specialist juvenile justice team. She had worked with Laura, an adolescent girl in care, for about a year on issues of petty offending. As their relationship strengthened, Laura began to tell Frances about the severe abuse and exploitation that she had suffered in her family of origin. At the same time as this, Laura's offences of theft (from shops) and violence (towards other adolescents) became more serious. The pressure from the juvenile justice system, and from her own system, to control this young person was very strong – even though this would mean her moving 'up the tariff' of possible court disposals. On the other hand, Frances's own training and influence from her colleagues who specialised in child protection led her to feel that, at this stage, Laura needed to feel cared for, rather than controlled.

Frances felt pulled in two opposite directions, with each 'camp' very critical of the other. Frances already felt very stressed by the task that had been given to her, but felt even more pressurised by the impossibility of resolving the two conflicting expectations of task outcome. (Thompson *et al.*, 1994)

Resistance and intractability

The central belief around which social welfare has developed is a belief in the possibility of positive change in clients and their circumstances. Unfortunately, because of resource problems, client and societal resistance and intractability, this positive change is often difficult to achieve. Because social welfare work is characterised by workers having high expectations of self, this intractability or inability to achieve change can become reframed as the worker's own fault, or caused by their lack of commitment and/or professional skill.

Violence and aggression

Social welfare work is often located in places of conflict – generally conflict within the family, or conflict between the state and the family or individual. Violence can often result. Furthermore, social welfare staff have to work in close proximity with clients whose problems often include uncontrolled aggression and violence.

PRACTICE FOCUS 1.10

Dee was an older and very experienced residential social worker, who was proud of her ability to make relationships and 'cool down' even the most aggressive of adolescents. After a very minor altercation with Karen, a fifteen year old that she was quite close to, the child got drunk, boasted of her intended course of action, returned and was systematically violent to an extremely upset Dee.

Social welfare workers often find themselves in teams that have a 'macho', be tough culture. In these teams the likelihood of violence may be ignored or underestimated in a spirit of bravado, leaving the negative effects of violence to be swept under the carpet. A popular myth surrounding violence and social welfare work is that a practitioner who has been assaulted somehow becomes immune to the effects of violence. In reality, the opposite is far more likely to be the case – the more one has been the victim of violence, the more debilitating the threat of anxiety around violence is likely to become (Jones *et al.*, 1991).

We have no clear information about why hundreds of social welfare workers quit the social welfare arena each year. However, practice experience indicates that a significant number will leave after an assault. That is, violence can be seen to hurt at both a physical and a psychological level, perhaps terrorising us, taking away our power and control over our lives. It can disorient us, leaving us feeling uncertain about our environment. Perhaps worst of all, we often develop a sense of responsibility for that violence, telling ourselves that it is understandable and, if only we had acted or responded differently, it would not have happened. This is a point to which we shall return below.

PRACTICE FOCUS 1.11

Maureen was a field social worker who had recently been assaulted whilst visiting a couple who misused drugs. A neurological examination of Maureen revealed that the use of her right arm and leg was impeded in that she had lost the sensation of pressure in both, though she could move them voluntarily. The neurologist was unsure how quickly, if at all, the damage would be repaired. Maureen was prescribed daily physiotherapy, and had to wear a support collar for the damaged neck. In addition part of her head continued to feel numb. Maureen was referred to the counsellor four weeks after the assault. She had constant flashbacks to the day of the assault. She was reliving the whole scene as if it were conducted in slow motion, starting with the baby's blank gaze, the sight of Matt sprawled out on the settee, and Jane picking up the hairbrush and advancing towards her.

Not only does Maureen relive the scene of stress as though it were a slow-motion nightmare, but she finds her physical debility a constant reminder of the intrusion that she suffered, which causes her stress many weeks after the event.

(from Scott and Stradling, 1992)

The politics of social welfare

Social welfare work can often involve a tension between the needs of individual clients/service users or families on the one hand and the expectations of the State or broader society on the other. This is because social welfare is intrinsically a political activity in the sense that it necessarily involves the use of power and relates very closely to the way society is structured and organised (Thompson, 1992). Social welfare workers are also publicly accountable for their actions and are answerable directly or indirectly to local or central government.

This political basis of social welfare brings with it additional pressures in terms of tensions, conflicts, dilemmas and the constant danger of unfavourable media or public attention. While social welfare workers are by no means the only professional group to face such pressures, the difficulties are very much part and parcel of social welfare.

■ The negative impact of stress

In addition to recognising the pressures inherent in social welfare work, we also need to acknowledge the immensely destructive effects such pressures can have if they are not dealt with effectively or appropriately. It is therefore important to consider the negative impact that stress can have.

Social welfare relies on individuals and teams investing in, and seeking to improve, the lot of individuals, families and groups. In order to be able to make this investment, a high degree of motivation, skill and energy is necessary. Stress reduces motivation, energy and sense of skill. In effect, by debilitating staff, stress debilitates the social welfare service overall, as it is largely through the direct efforts of staff that results are achieved.

Frequently, in our often over-stretched services, there is a temptation to keep on feeding extra work to an already stressed and depleted workforce (the 'willing donkey' syndrome). In the end this tends to be counterproductive as the quality (if not the quantity) of work will almost always be substantially reduced. A failure to recognise the significance of stress and its harmful effects can therefore contribute to:

- lower standards of practice and service delivery
- a greater risk of mistakes being made
- an unhappy, unhealthy, unfulfilled workforce, and
- a fraught or even oppressive working environment.

A further possible consequence is that of 'burnout', an important concept that merits closer attention.

■ Burnout

Burnout is a psychological condition that can arise as a consequence of exposure to high levels of stress over an extended period of time. It is a highly problematic condition that can do a great deal of harm to all concerned.

'Quick and Quick (1984) note that stress can cause a number of indirect costs. These include loss of vitality, as well as the communication breakdown and faulty decision making referred to above. Loss of vitality may lead to low morale, low motivation and high dissatisfaction. Ultimately this will lead to burnout and to poor provision of services. Burnout is a condition closely linked to stress and is characterised by three distinct but related phenomena (see Maslach and Jackson, 1981):

- *emotional exhaustion*
- *lack of individual achievement*
- *depersonalisation.*

The overall effect of burnout is a very destructive one, leaving the staff concerned feeling bitter, disillusioned and demotivated – and therefore unlikely to achieve high standards of professional practice. They become locked into a cycle of negativism in which their morale and well-being suffer, as do the effectiveness of the organisation, the interests of service users and the profession of social work as a whole.'
(Thompson *et al*, 1994, p. 13)

Burnout is a condition that can be avoided if appropriate steps are taken to protect staff from long-term exposure to stress. However, if such measures are not taken, if staff are left to struggle with the sometimes immense pressures they face over an extended period of time without relief or adequate support, then it is not surprising that burnout is so often the result.

In effect, burnout can be seen as an attempted solution to the problem of stress. If the pressures cannot be abated, removed or controlled, then we may be left with little choice but to change our attitude towards them, to 'detach' ourselves from the emotional reality of the situations we face, hence the 'depersonalisation' and lack of emotional engagement associated with burnout.

What is stress?

Pressure and stress

Stress is a term which is used in a number of different ways, both in general conversation and in the research literature. It is therefore necessary to be clear in what sense we are using the word, to be explicit about how it is defined. A helpful definition – and the one informing the ideas in this manual – is that of Arroba and James (1987): 'Stress is the individual's response to an inappropriate level of pressure. It is the response to pressure, not the pressure itself' (p. 3).

Pressure is a demand made on our time or energy and, as such, it is 'neutral' – that is, it is neither inherently problematic nor inherently positive. Whether pressure is experienced as good or bad depends partly on the amount of pressure and partly on our appraisal of it and, hence, our response to it. Where pressure is positive it is a source of motivation and stimulation. Where it is negative, it is experienced as stress – that is, it is harmful.

The level of pressure can be inappropriate in two ways:

1. Too little pressure – boredom, apathy and a lack of commitment are the likely result;
2. Too much pressure – staff who are overburdened are more likely to make mistakes, fall ill, and so on.

The task of all concerned with dealing effectively with stress, therefore, is to help to ensure, as far as possible, that staff remain within the optimal middle range of pressure, where it is not too little to motivate and not too much to overwhelm. Managing the flow of pressure is therefore a central part of stress management.

The three dimensions of stress

Stress can be seen to have three dimensions or elements:

- stressors – those factors which produce pressure and are therefore potential sources of stress;
- coping methods – the skills and strategies used to deal with pressure and manage stress;
- support systems – the range of possible sources of support, including formal support (work-based support systems) and informal (partner, friends, colleagues, and so on).

These three elements are often oversimplified to two – stress is then seen in terms of a struggle between stressors (which threaten to overwhelm the individual) and coping methods (which are used to resist the pressure). What this picture neglects is the significant role of support which has a major bearing on both stressors and coping. A well-supported person is less likely to experience pressures as stressful, and good support also enhances coping methods, for example, by boosting confidence. This relates back to the definition of stress above – one's response to an inappropriate level of pressure. Support systems can help to keep the level of pressure within the optimal middle range and can help to ensure that the worker's response to the pressures is a positive and confident one. Clearly, then, the role of the manager is a pivotal one in the management of stress.

Ellen was going through a particularly difficult period in her caseload – everything seemed to be 'blowing up' all at once, crisis following crisis. In the middle of this Ellen experienced Rashida, her manager, as being very supportive. Rashida managed to take small pieces of work from Ellen, took several difficult phone calls and most importantly, was always available for informal consultation.

Myths about stress

A number of myths about stress abound and we do not have the time or space to explore all of them in full. However, it is worth noting some of the more damaging ones. For each of these, space is provided for any notes you may wish to make about your own views, experiences or examples.

Myths about stress – 1

A true professional never experiences stress or, at least, never gives in to stress.

This is one of the most powerful myths around being a professional. It seems that being a professional dehumanises the individual and gives them invincibility against stress.

Notes:

Myths about stress – II

A true professional never lets home stress affect his or her work performance.

This myth assumes that the different stresses can be picked up or put down at the end of the particular home or work shift. In reality, stress does not recognise these boundaries, crossing them at will.

Notes:

Myths about stress – III

There are others much worse off than you (stop whinging).

This myth is a particularly powerful one for social welfare staff who work with particularly disadvantaged people every day. The practitioner is always working with someone 'worse off' than themselves and must therefore be relatively stress-free. This myth denies the intrinsically personal experience of stress which makes such comparisons meaningless.

Notes:

Myths about stress – IV

Staff only get stressed by specific, major crises.

This presumes that staff will only feel stressed following their involvement in a major trauma or crisis. In reality severe stress can arise from a combination of accumulating smaller stressors and the gradual undermining of individual coping mechanisms (see Practice Focus 1.13 on page 28).

Notes:

Myths about stress – V

Staff get stressed but managers can take the pressure (if you can't stand the heat...).

Most unhelpful of all for stressed managers is the presumption that they are immune from stress, and (particularly in macho managerial cultures) able to cope with anything. Social welfare managers experience stress daily: this myth can hold them back from admitting it and getting help. (See Practice Focus 1.14 on page 28).

Notes:

Myths about stress – VI

Stress is good for you.

This arises from differences in terminology. Some people do not distinguish between pressure and stress and therefore see stress as either positive or negative, depending on the circumstances, with 'positive stress' being a challenge that some will rise to. We must be wary, then, of allowing confusion over terminology to mask the serious damage caused by stress as opposed to the motivating force ('butt-kicking') that many find in a challenge.

Notes:

Myths about stress – VII

Stress is a sign of weakness.

This reflects the oversimplified, two-dimensional model of stress mentioned above, seeing stress as a personal failure to cope with pressure. This ignores the crucial role of support and thereby 'blames the victim' (see page 39).

Notes:

Myths about stress – VIII

Stress is experienced by managers only.

This is the notion of 'executive stress', and is based on the mistaken idea that 'ordinary' workers do not experience stress. Managers and staff may well experience different types of stress but we should not allow this to mislead us into assuming that stress is the preserve of managers (though having responsibility for the management and well-being of others can be a particularly taxing stressor – see Part 3).

Notes:

Myths about stress – IX

Hold your breath and everything will be OK.

A view commonly held by social welfare staff, both managers and practitioners, is that as long as you grit your teeth, hold your breath and hope for better weather, stressful times can be survived. This is called the 'fingers crossed' technique. While all social welfare staff may use this technique from time to time, it can often prove a very unhelpful response in so far as it can prevent us from addressing the sources of stress.

Notes:

Myths about stress – X

I'm not that kind of person.

A further idea that is often encountered in social welfare work is that some staff are the right 'type' to feel stressed and need to draw on staff care facilities, but others are not. This notion that 'it only happens to other people' is a dangerous one as it means that we may not take seriously the stress factors that we face or we may not seek help when we need it.

Notes:

PRACTICE FOCUS 1.13

Maria was a very organised worker, with systems and a good method of prioritising her work. During one particular period in March, five clients developed different but serious problems, including a substance overdose, disclosure of abuse and an unexpected family bereavement. Maria's system fell apart – she felt very stressed. In this case, skilful intervention by her manager allowed her to work through this situation, helping her to change the personal meaning she attached to them (in particular realising that they did not reflect on her ability as a worker and were not her fault).

PRACTICE FOCUS 1.14

Melvin managed the busy intake team. He was respected by his colleagues and very sensitive to other people's stress, sometimes helping them to make the first contact with the staff care scheme. The year had not been good for Melvin. He had gone through an acrimonious separation and had had particular problems with a member of staff who was really struggling to make the grade. At no time did Melvin talk about his stress with his manager. At no time did he consider using the staff care scheme himself. He was convinced that he could 'tough it out' until a collapse at work brought matters to a head.

■ Sources of stress

Potentially, the number of sources of stress is infinite but there are some common patterns that can be identified. These include:

- work – too little or too much work can be stressful, as can work which is too difficult or demanding (for example, dealing with bereavement);
- roles – uncertainty about what role one is expected to play or conflict between differing roles can be major sources of stress;
- strained relationships – these may arise as a result of personal or work-based factors, or from a mixture of the two;
- organisational culture – a negative or strained atmosphere can produce or exacerbate feelings of stress;
- home-work interface – pressures at home and at work can combine to produce an intolerable burden, particularly if work arrangements (for example, working overtime) interfere with one's home life;
- critical events – these are events which are associated with major trauma, change or loss. At home they can include birth, bereavement, divorce or relocation. At work, two of the most significant include:
  - reorganisation/relocation: these significant changes in the workplace are often harbingers of stress, especially when they are not very well-managed;
  - violence in the workplace: is a particularly potent source of stress to its recipients;
- emotional pressures – see the discussion of 'emotional labour' below;
- the politics of welfare, as discussed above.

There is, of course, no point in trying to develop a definitive list of sources of stress. What is needed in its place is a sensitivity to stress factors, the ability to recognise, and ideally anticipate, a staff member's vulnerability to stress. This is an important point and one to which we shall return below. However, there are two particular elements from the list above that are worthy of further, more detailed comment as we see them as particularly pertinent to social welfare work and the context in which it takes place. These are violence/aggression and 'emotional labour'.

Aggression and violence

As already noted, aggression and violence can be very significant issues in social welfare and, by extension, very significant sources of stress. It is important to note that they can prove to be stressful in a number of ways:

- **Fear of violence**
  At times, our own anxieties about the potential for aggression and violence can produce high levels of pressure, even where those anxieties are relatively unfounded. This illustrates that pressure and stress have a very strong subjective dimension in which our *perception* of the situation is extremely important, even on those occasions where our perceptions are inaccurate or mistaken.
- **The escalation from aggression to violence**
  An aggressive response does not necessarily lead to an act of physical violence. However, when someone responds aggressively to a particular situation, the

social welfare worker is likely to experience a great deal of pressure due to the fear that his or her comments or actions may lead to an escalation of the situation, resulting in an actual assault.

- **Self-protection**
  Where incidents do occur, the impact on workers can be immense in terms of the pressures to protect oneself from harm and to protect vulnerable others as well. Such incidents are usually very fraught and therefore potentially very stressful indeed.

- **Feelings of guilt**
  A common response to incidents of aggression and violence is a strong sense of guilt. This is very similar to responses to loss where irrational feelings of guilt are commonly experienced. People often wonder: 'What could I have done to prevent that happening?', 'What did I do that caused that to happen?' or 'If only I had ...'. It is important to recognise that, although these feelings are often *irrational* (that is, they are unfounded), they are nonetheless very real feelings and therefore a significant source of pressure. It can be very harmful indeed if people try to dismiss such feelings or regard them as unimportant.

- **The aftermath**
  The way the 'aftermath' of an incident of violence or aggression is handled can be crucial in terms of the experience of stress. Such incidents tend to generate strong feelings, feelings that can prove very debilitating if they are not dealt with appropriately. If the aftermath is handled insensitively by the organisation or by specific colleagues, the harmful effects can be exacerbated. It is therefore vitally important that opportunities for debriefing or other supportive measures are made available where required.

- **The next time**
  Subsequent to an incident of aggression or violence there can be problems and pressures that arise from the next time the worker deals with the individual concerned or has to handle a situation which has much in common with the incident. The 'next time' can generate high levels of anxiety and is likely to need handling very sensitively and supportively.

The question of aggression and violence is therefore a very complex one with far-reaching implications. It therefore needs to be given a great deal of thought so that the pressures that arise are managed as effectively as possible and their potential negative effects minimised.

Emotional labour

Hochschild (1983) used the concept of 'emotional labour' to refer to the process of having to manage one's emotions for the benefit of effective interactions with service users. The 'emotional management' that we routinely undertake as part of our everyday life has to go a stage further for workers in social welfare and related disciplines. As such, emotional labour has the potential to be a further source of job-related stress in so far as it can increase the demands made of the worker. This is a very complex area of psychological study with a number of factors that have to be taken into account (see Wharton, 1993). However, there are a number of conclusions we can draw:

- Social welfare involves emotional as well as physical and intellectual demands. A sensitivity to the emotional dimension of the work is therefore called for. A macho, be tough attitude is unlikely to be helpful and can prove very harmful (Pottage and Evans, 1992).
- There is a possibility of 'emotional dissonance' (Ashforth and Humphrey, 1993). This refers to the ways in which what we feel may differ from what our professional persona is expected to demonstrate. For example, in dealing with a man who has sexually abused a child, we may feel considerable revulsion towards that person and yet still need to be able to work constructively with him. (It can take considerable skill to manage this conflict, for example by learning to express revulsion towards what he *did* rather than towards what he is).
- The extent to which we are directed by others in terms of what we should be **feeling** can be a further source of pressure. For example, if managers or others attempt to impose some degree of control over our emotions, we may find this difficult to deal with, and so additional stress may be the outcome. It is therefore important that managers do not try to dictate to staff how they should be feeling.

■ The times they are a changin'

One very important aspect of stress that cannot readily be ignored is the central role of change. Changes can be positive or negative, good news or bad, but always bring additional pressures. Consider, for example, the case of 'spend, spend, spend' pools winners whose lives have been ruined by not being able to adjust to the pressures of the major changes in their lives.

Change brings an element of threat – existing strengths may no longer apply, new weaknesses may appear and the whole situation can be characterised by strong feelings of threat. Of course, change does not automatically lead to stress but we do need to recognise that it is a significant risk factor:

1. **New stressors**
   Change always brings new work pressures that can become stressors.
2. **Loss of old coping mechanisms**
   Change frequently leads to individual coping mechanisms being lost or left behind. This leads to an increased vulnerability to stress.
3. **Loss of staff/agency support**
   Change can lead to appropriate staff care being unavailable/unknown to the member of staff.

This is particularly significant in view of the rate and range of changes staff constantly face. As people are often heard to say 'The only thing that's stable these days is change'. In view of this, change needs to be taken seriously and handled sensitively. The climate of change makes it all the more important that we are as well-equipped as possible to meet the challenge of stress.

The climate of change in social welfare

As Pottage and Evans (1992) point out, not only has change become endemic to social welfare agencies, but this change is usually concentrated on the need to squeeze more out of existing staff and resources. As our own study has shown (Thompson *et al.*, 1996), one of the hidden costs in wholesale change and reorganisation can be the significant debilitation, through stress, of a whole agency or significant group of staff. There are, therefore, important steps that need to be taken if the pressures associated with change are not to be allowed to overwhelm people.

The climate of change in social welfare arises as a result of the coming together of a number of factors, not least the following:

- **Political and ideological changes**
  The debate about society's duties towards its citizens has been a major feature of political discourse in recent years. As part of this, there have been significant moves to reshape (or even dismantle) the welfare state. The change brought about by political and ideological influences has therefore been of major proportions.
- **Legislative changes**
  Largely but not exclusively as a result of the political and ideological changes, there have also been a number of major changes in the law and statutory guidance relating to social welfare work. These have, in turn, led to major changes in policy, procedures and practice.
- **Organisational changes**
  There have been two sets of organisational changes that have played an immense role in contributing to an overall climate of change and instability: Firstly, local government reorganisation and the move to unitary authorities; and secondly, a preference, stemming from current management theory, for organisational restructuring as a means of attempting to resolve organisational difficulties and maximise efficiency and effectiveness.
- **New technology**
  The increasing use of computer technology in social welfare organisations has necessitated a rethink of a number of aspects of working life. Computers are not only a basic element of record-keeping and data storage but are increasingly being used as forms of communication or even as practice tools in their own right (Bates and Pugh, 1995).

Clearly, then, the question of managing change is one that needs to be given very serious consideration if stress is not to be allowed to cause major problems for staff, their employers and the users of services.

Meeting the challenge

It is our contention that everyone has a responsibility to play a part in seeking to meet the stress challenge. Everyone suffers as a result of stress and so it is important that everyone makes a contribution to ensuring that stress levels are kept to a minimum. However, we should note that managers have a particularly strong duty to make such a contribution as the wellbeing and effectiveness of staff is a significant feature of their duties. We therefore concentrate specifically on the concerns of managers. Issues relating to training and staff development are addressed in Part 2 and to practitioners in Part 4.

■ The manager's role – prevention, response, recovery

The manager's role in meeting the stress challenge can be seen to be threefold: prevention, response and recovery.

Prevention

This involves:

- Understanding the causes of stress and trying to make sure these are kept to a minimum within the work environment. For example, a lack of clarity about roles may need to be addressed.
- Creating a positive atmosphere in which staff feel supported, valued and able to discuss pressures and problems openly.
- Recognising the signs of stress (or impending stress) so that problems can be 'nipped in the bud' wherever possible.

At the case allocation meeting, the Faulkner family seemed ideally suited to Siobhain, but she was strangely reluctant to take them on board. Her manager noticed this and that Siobhain's mood began to alter in the following weeks. One month later, in supervision, Siobhain's manager told her that she was concerned about her and the case. Siobhain disclosed that the Faulkners brought up distressing echoes from her own childhood. Following this supervision, in a very discreet fashion, the work was reorganised and Siobhain chose to go to talk to someone about those childhood issues.

Another important aspect of prevention from a managerial point of view is the need to develop a policy on dealing with stress in particular and developing staff care in general. We return to this point below under the heading of 'staff care'.

Response

Preventative measures, whilst very worthwhile in their own right, cannot be guaranteed to be effective in every case. There is, therefore, a need to have support measures and systems to help people through a period of stress – to make a positive and constructive response. These include:

- Opportunities for discussion, 'debriefing', or 'de-responsibilitising' (see below) especially after a traumatic incident, for example, where someone is involved in an accident or has been assaulted.
- Help to prioritise or re-allocate work tasks to create a breathing space and to help the organisation and the individual reduce what is expected of them.
- Making a referral for specialist confidential counselling if appropriate.

Dee's manager had realised the impact of Karen's assault upon Dee. She debriefed her, reassuring her that she was not to blame for the assault, offered her choices about what she did next and made an early referral to the staff care scheme.

Recovery

After a period of stress, help is likely to be needed to 'recover', that is to readjust. Unfortunately, this is a neglected area of stress management and a lot of mistakes are made, for example, expecting people to 'get over it' too soon. At this stage, then, the manager's tasks include:

- Again, offering opportunities for debriefing or possibly confidential counselling.
- Helping to plan a gradual reintroduction to work patterns.
- Helping to guard against feelings of self-blame and low self-esteem, for example, through praise and encouragement.
- Helping to protect the staff member from a speedy repetition of the stress situation.
- If the member of staff is away on sick leave, maintaining the delicate balance between keeping contact to prevent isolation and being intrusive.

One very important point to recognise is that the role of the manager is a pivotal one – it can, and often does, make a crucial difference as to whether and how stress is experienced.

'If a manager is not helping, then he or she is surely hindering, as a lack of formal support is likely to be interpreted in terms of the following:

- *a lack of security or 'safety-net'*
- *a feeling of being undervalued*
- *a generally low level of morale*
- *possible resentment of managers felt to be 'not earning their pay'.*

When the actions or attitudes of managers contribute to producing such a situation, the level of pressure is increased, rather than decreased.'
(Thompson *et al.*, 1994, pp. 32-3).

So, as Arroba and James put it: 'Managing pressure and reducing stress is not an optional extra for managers' (1987, p. xvi).

■ Recognising stress in others

Different people respond to stress in different ways and so it is not possible to give a definitive guide to recognising the signs of stress (and see Section 3). However, although stress is experienced in an individual way, its symptoms are held in common – 'tell-tale' signs include:

- increased anxiety
- constant tiredness
- insomnia/oversleeping
- apathy/helplessness
- irritability
- a tendency to make more mistakes than usual
- increased blaming of self or others
- absent-mindedness
- intrusive tearfulness
- obvious discomfort with normal team dynamics.

The list could be a very long one and, of course, many of these 'signs' occur on an everyday basis. What we need to be sensitive to is a pattern of signs, rather than simply one or two isolated indications. Indeed, this is a central point – a 'checklist' of signs of stress is no substitute for genuine sensitivity. Each staff member has his or her own 'band' of normal behaviour and mood – thus some people externalise their internal dialogue ('have a good moan') as part of their normal way of coping with taxing demands, whilst others may do this only when they are beyond coping – so it is when a staff member strays outside his or her normal band in a significant way that something is not quite right.

■ 25 ways to stress out your staff!

As we noted above, the manager's role is a central one in so far as his or her attitudes and actions can either help considerably or make matters much worse. It is therefore important that managers are aware of the possible ways in which they can contribute negatively to stress management – that is, appreciate how they can so easily be part of the problem, rather than part of the solution.

Listed in the following pages are twenty-five possible ways in which a manager can hinder rather than help. These are provided as a means of helping to develop an understanding of, and sensitivity to, some of the complexities of dealing with stress. As with the myths surrounding stress that were discussed earlier, space is provided for you to make notes about your own views, experience or examples.

1. Minimising the problem

This is what happens when someone tries to play down the importance of a problem, feeling or situation. The difficulty with this approach is that it tends to do more harm than good. The reason for this is that stress has a subjective dimension – stress is our response to an inappropriate level of pressure. Consequently, a tendency to minimise the problem is likely to have the effect of making people feel that their concerns are not being taken seriously or their feelings are unimportant. Clearly this will result in more pressure, rather than less, and a feeling of being unsupported and therefore vulnerable. It is thus essential that managers are able and willing to listen – to take their staff's feelings and concerns seriously.

Notes:

2. Management by exception

This term refers to the management style which results in staff only seeing a manager when something goes wrong. If everything is running smoothly managers are rarely seen, but they soon appear if mistakes are made or difficulties encountered. This creates unnecessary barriers between managers and staff and leaves the latter feeling unsupported – managers come to be associated with scrutiny rather than support. It is therefore important that managers are seen as part of the team, rather than just a potential threat to it.

Notes:

3. Blaming the victim

This refers to the process in which the victims of certain circumstances come to be seen as causing the problem, rather than suffering from it. For example, poverty is seen by some people simply as the result of laziness on the part of the poor without taking any account of the economic or social factors underpinning poverty. Similarly, people experiencing stress are often seen as 'weak' or 'inadequate', without sufficient account being taken of broader issues of levels of pressure and the quality and quantity of support available. Managers therefore need to be wary of adopting too narrow a perspective and thereby falling into the destructive trap of 'blaming the victim'.

Notes:

4. Guilty until proven innocent

A not uncommon complaint from staff in some areas of work is that their confidence is often undermined by the apparent assumption, on the part of managers, that they are not to be trusted. That is, it appears that they are assumed to be incompetent unless and until they prove themselves to be otherwise. This lack of trust undermines both confidence and professional autonomy. It can also prove to be a self-fulfilling prophesy – a person whose confidence is undermined by a lack of trust is more likely to produce poor quality work. By the same token, a person who feels valued and trusted is more likely to have higher self-esteem and produce better work.

Notes:

5. Getting on with the job

This is a similar approach to 'minimising the problem' mentioned above. It involves being so anxious to move on and get the job done that we fail to notice the pressures that staff face. The danger here is that the anxiety to get the job done can actually make it less likely that it will get done – due to the unnecessary strain placed on key staff. It has long been recognised by successful organisations that driving staff as hard as you can is not the way to get the best out of them. Indeed, it is a good way of increasing stress and thereby decreasing effectiveness. The 'tunnel vision' of getting on with the job is therefore to be avoided.

Notes:

6. Blocking job satisfaction

Job satisfaction can be seen as the converse of stress – it is where pressure is in the optimal middle range and acts as a source of stimulation and motivation. Job satisfaction is also an important safeguard against stress. If staff are getting a lot of positives out of their work, they will be more resilient and better equipped to deal with the pressures. It is therefore important that managers do not stand in the way of job satisfaction, for example by prioritising work tasks in such a way that there is no let up in the more demanding and less enjoyable or fulfilling tasks. If staff are not to be worn down, then the issue of job satisfaction is one that needs to be taken very seriously.

Notes:

7. Favouritism

It is inevitable that we will like some people more than others and so there is always the danger that we will have 'favourites'. For those staff who are less frequently favoured, this can be an additional source of pressure and may also act as a barrier to seeking support. Ironically, this can also be seen to apply to those who are the 'favoured few'. They may feel uncomfortable and not welcome, being separated from their colleagues by barriers of jealousy, resentment or underlying conflict. In short, favouritism does nobody any favours. Consequently, it is important to ensure that all staff are treated fairly, regardless of how much we like or dislike them. The task is to support all staff, not just the ones we like.

Notes:

8. Tokenism

This describes the tendency to 'go through the motions' without really being committed to genuine change. This is a strategy some people adopt deliberately but many other people slide into this type of behaviour unintentionally, by default as it were. They 'make the right noises' but do not actively pursue solving the problem concerned. When this is applied to stress management, the situation quickly becomes critical. A manager who offers token support without a genuine basis underpinning it runs the risk of alienating staff quite considerably. Tokenism is very easy to spot and it causes immense resentment and ill-feeling. Commitment to managing stress therefore has to be more than skin-deep.

Notes:

9. Taking people for granted

Managers are busy people with many competing demands upon their time and energy. It is understandable, then, that many things will not be given full attention and will be taken for granted. When this applies to people, though, the results can be disastrous. As we emphasise throughout, people are the most important part of any service organisation – everything is achieved through people. We need to recognise, then, the key role of people in achieving success. In particular, we need to be careful not to take for granted the more competent staff – to assume that 'good' staff do not need support. Support is both an energiser and safety net for all staff.

Notes:

10. Ignoring critical incidents

A critical incident is one in which a great deal of emotion is stirred, for example, a loss, an accident, an attack (physical or verbal) or an embarrassing situation. Dealing with staff who encounter such an incident can make us feel very uncomfortable and can stir up our own emotions, thereby making us feel vulnerable. A common response to such situations, then, is to avoid them, to play down their significance and expect staff to 'pull themselves together' and quickly get back to normal again. This is potentially disastrous. Critical incidents which are not dealt with sensitively, constructively and fully can cause considerable harm, both in the short and long term.

Notes:

11. One-way pressure

Modern organisations can be characterised by an 'overload' of pressure, too much to be done by too few people – and all within a context of constant change, a constant flow of new demands and new challenges. This means that the excess pressure in the system has got to go somewhere. The danger is that, because of the way power works, the pressure is likely to go downwards, down to the least powerful members of the organisation (and ultimately down to the clients or customers of the organisation). However, if an organisation is to be effective, some of that pressure has to go upwards, up to the people with the power to do something about the problems. This reflects a key aspect of the line manager's role – 'fighting up and selling down'. If it is all selling down and no fighting up, the organisation will not be a responsive one and the likelihood of staff being crushed by stress will be increased.

Notes:

12. Sexual harassment

This is a subject which causes considerable ill-feeling and can do great harm – and yet it is often dismissed with a joke or trivialised. Unwanted sexual attentions are a major source of pressure and can easily overshadow other workday concerns for the person on the receiving end. There are two very important sets of issues for managers: commission and collusion. Managers should clearly not commit sexual harassment through unwanted touches, sexual comments and so on, even if these are intended as a joke – what appears as a joke to a person in power can appear as a form of subtle oppression for a less powerful employee. Managers also should not collude with sexual harassment by contributing to, or tolerating, a work environment in which women are portrayed as sexual objects. In a social welfare service where, in the main, women deliver and men manage, the potential for sexual harassment is high. Accomplished 'harassers' become adept at reframing or concealing their behaviour and choosing as targets those who are most vulnerable.

Notes:

13. Racial harassment

Much of what has been said in relation to sexual harassment can also be seen to apply to the harassment of racial or ethnic groups. In line with anti-discrimination legislation and humanitarian moral values, managers have a duty to ensure that staff do not receive unfair or unfavourable treatment on the grounds of the colour of their skin, their ethnic background, their religion or their language and so on. Where such discrimination is allowed to go unchecked, certain members of staff can experience considerable stress as a result of the hostility, lack of respect and undermining of confidence they encounter.

Notes:

14. Allowing discontent to fester

This refers to the tendency not to tackle difficult issues and simply hope they will go away. Where there is conflict or discontent, a 'head in the sand' attitude is highly unlikely to prove effective. In fact, it is more likely to have the opposite effect as discontent which is allowed to fester can very insidiously undermine morale and good will and, ultimately, cause good staff to leave. It is therefore vitally important that matters of discontent are tackled head on, clearly, openly and constructively. This not only prevents 'festering' but also creates an atmosphere of trust, security and, thereby, support.

Notes:

15. Brushing problems under the carpet

This has parallels with No 14. It entails taking a passive approach to problem-solving and can leave staff feeling anxious that nothing is going to be done about important problems and concerns. Often, this situation arises because the manager is under too much pressure to deal with that particular problem. If this is indeed the case, two issues arise: first, if there is too much pressure in 'the system', this should go upwards, rather than downwards (see No 11) so that staff are not overloaded. Second, if managers cannot address particular problems themselves, then consideration should be given to delegating responsibility, rather than abdicating it by brushing the problem under the carpet.

Notes:

16. Betraying confidences

It ought to go without saying that managers should not betray confidences, and yet it is not uncommon for staff to complain that this is precisely what has happened. A supportive managerial relationship needs, of course, to be premised on trust, and so the betrayal of that trust can have major consequences and leave the staff member concerned feeling vulnerable and demoralised, as well as let down. It is therefore important to be clear in dealings with staff what is to be treated in confidence and what is not.

Notes:

17. The macho approach

'Macho' management has been heavily criticised in recent years. Staff who comply on the basis of fear are unlikely to fulfil their potential or work to maximum effectiveness. Such an approach encourages defensiveness and breeds resentment and mistrust. A more participative management style which encourages staff to take ownership of their work is far more likely to be effective and will create a supportive atmosphere of openness and collaboration, rather than one of fear and loathing (Harding, 1987). It needs to be recognised that a 'macho' bullying style is a sign of weakness, not strength. This is particularly the case with regard to stress where a 'be tough' attitude can be so destructive (Pottage and Evans, 1992). The macho approach denies the existence and importance of stress for manager and staff member. Where stress is not acknowledged it cannot be recognised and dealt with. Because social welfare is largely managed by men and delivered by women, the macho approach can be used to deny the feelings, needs and perspectives of women workers.

Notes:

18. Ignoring the work environment

A common source of stress is the physical environment of the workplace. Poor lighting, heating, ventilation, lack of space or privacy, intrusive noise, inadequate facilities all play a part in increasing pressure and allowing stress to flourish. Although managers may have little control over the environment, it is none the less important that any possible steps are made to make the physical environment as 'user-friendly' as possible. This may be through making small-scale changes or adjustments, or through making representations to senior management for improvements to be made and deficiencies rectified.

Notes:

19. Being a 'stress-carrier'

A 'stress-carrier' is someone who is anxious and overwrought and, through his or her dealings with others, transfers the anxiety to them. Such a person 'winds people up' simply by behaving in a tense or agitated way. Such behaviour is unnerving and unsettling and undermines any sense of security or support the manager should be trying to instil. Managers who are prone to this therefore have to learn to relax and adopt a more controlled approach to their work. If not, 'stress begets stress'.

Notes:

20. Not saying thank you

So far as stress management is concerned, a very important saying is: 'Little things count for a lot'. A simple matter like a manager not saying 'thank you' can often be enough to cause resentment and feelings of not being valued or appreciated – feelings which undermine morale and resilience to stress. It is a mistake for managers to see such matters as trivial, as this shows a failure to appreciate how significant our day-to-day interactions with staff are – the significance of the 'messages' managers give to staff. In short, a little appreciation goes a long way.

Notes:

21. Stick to your role

Within social welfare settings, staff members will frequently adopt a specific informal role – joker, social secretary, workhorse, cynic, carer, leader of the opposition, are some of the more common ones. At times of stress, staff can find these roles become oppressive or impossible to maintain. Unfortunately, the team and manager frequently want the person to continue in the role. A positive approach will lead the manager to insist on the staff member 'taking a break' from playing their role during their period of stress.

Notes:

22. It's all your fault

After an assault or crisis, a manager may blame the staff member for everything that has occurred: 'If you hadn't done that, it would have all been all right' or 'If you hadn't said that, they wouldn't have become angry and you wouldn't have been assaulted'. In reality, although anger is a legitimate feeling and emotional response, violence is not a legitimate behaviour and is always the responsibility of the perpetrator, never the responsibility of the recipient.

Notes:

23. Standard-setting in supervision

There are three main component parts of supervision:

- standard-setting (Are you doing what you are supposed to be doing, and are you doing it well enough?);
- staff development (What do you need to learn? How do you need to develop your skills?);
- staff care (How are you? How is the work affecting you as a person?).

Especially within a macho management culture, the standard-setting element is often exaggerated to the point where staff development and staff care are completely ignored. Staff feel under more pressure, and the elements of supervision that might help to deal with this pressure are ignored.

Notes:

24. Absence does not make the heart grow fonder

Some managers rarely spend time with their staff. They always seem to have other priorities, other pressures that are more important. No-one realistically expects a manager to be around all of the time, or even most of the time, but problems can arise when staff start to feel that they are unsupported, that they have no-one to talk to about any demands, pressures or worries.

Notes:

25. Be my friend

It is not uncommon for very good working relationships to develop between managers and staff. Often, these can become important friendships. While there is nothing inherently wrong in managers and staff being friends (but note the problems of favouritism discussed above), there is a danger that such friendships can be allowed to get in the way of an appropriate manager-staff relationship – it can 'muddy the waters'. Where managers lose sight of the need for appropriate professional relationships with colleagues, additional pressures can be generated.

Notes:

■ Staff care

A major theme underlying workplace stress is the need to recognise the importance of staff care as an essential part of good management practice. It is now increasingly recognised that an organisation's most important resource is its staff. There is much to be gained from ensuring staff needs are catered for and a great deal to be lost if the needs of staff are neglected or dismissed as unimportant. Staff care is therefore not a luxury, but a basic part of effective management.

These days, many organisations are making staff care a matter of formal policy. That is, they are developing explicit staff care policies by clarifying and strengthening existing support measures and by introducing new ones where necessary. It is important, for staff care to be successful, that it is presented in positive terms for all staff. It needs to be seen as a necessary support service for all staff, rather than 'casualty' service for those who experience stress. The continuing occupational health of a workforce relies as much on 'maintenance' as it does on 'repair'. A central part of this is the need to focus on positive aspects of occupational health such as job satisfaction and personal development so that stress is seen as a risk to be managed by all staff, rather than a sign of weakness in those who experience it. Staff care is therefore something to be actively and positively promoted, rather than kept in reserve in case it needs to be drawn upon.

In this way it might be helpful to integrate reactive policies (such as staff counselling schemes or other Employee Assistance Programmes) with more proactive staff care policies. These policies could include measures from the time the member of staff joined an organisation until the time they left.

Recruitment and selection

One of the simplest and most serious causes of work stress is that the person chosen does not 'fit' the job, or the job does not 'fit' the person. This leaves the new member of staff with considerable role stress throughout their employment.

Welcoming

The individual joining the team is a human being as well as a practitioner. The tone and style of their relationships at work will be positively affected by a good welcome and negatively affected by its absence.

Steven and Sabina were friends. The both joined the Barrowdale agency within a month of each other. Steven joining a team in the north of the district, Sabina one in the south. When they met three weeks later to compare notes, Steven was elated but Sabina was already feeling quite depressed. On his first day Steven had found a welcome card on his desk signed by all the team. During the course of the day all his colleagues had dropped by to introduce themselves and had arranged to take him out for lunch the day after. Sabina, on the other hand, had found no card, a desk full of other people's junk and colleagues that were strangely indifferent to her arrival. She went home that night feeling very deflated and anxious about what the future might hold.

Induction

A major form of initial stress is the lack of appropriate induction to the job. Induction is meant to inform the new practitioner about the work they have to do and how that fits into the wider service. This should include both the official rules about how services should be provided, and the unofficial rules that indicate 'how we do things here'.

One of the problems with induction is that the rules and 'norms' become second nature to long-established staff who do not realise how important it is to explain the basics of team life.

Vee was a new, nervous member of staff. She had been in the team for four days before a colleague sensed her frustration and asked her what the matter was. Vee, upset and feeling inadequate, explained that she did not know how to get an outside line on the phone.

Mentoring

Mentoring is the process of linking, for support and advice, less experienced practitioners with colleagues who have more experience and who are better established within the team. In some settings, this mentoring occurs naturally, in others a more formal scheme is advisable. The benefits of mentoring can be quite significant in helping staff to feel:

- welcome and valued
- supported and 'cared for'
- relatively protected from potential problems and hazards
- secure and comfortable
- part of a team or supportive staff group.

Supervision and appraisal

Supervision is the individual or group facilitation of good client care and good staff care. Supervision is a major player in individual and service development. Its three key elements are standard-setting, staff development and staff care, and all are essential in looking after staff. Appraisal is the yearly or bi-annual task of assessing individual development and future development needs. If undertaken in a positive fashion, appraisal can have a positive effect in terms of stress management.

Training

It has long been recognised that training for the complex tasks of social welfare is not a luxury but a necessity. It is reasonable to assume that training will address agency, team and individual training needs. Good training can be a very significant instrumental factor in reducing pressures and increasing people's abilities to cope effectively with stress. Conversely, poor training or a lack of training can have the effect of increasing it, partly by leaving staff feeling ill-equipped to deal with the demands that they face in their day-to-day practice and partly by increasing a sense of resentment and frustration about not being adequately supported.

Staff counselling

We would argue strongly that the provision of a staff counselling scheme is not enough in itself, as practitioners and managers will often be reluctant to use it. If the provision is seen as a normal investment in staff and the culture within the organisation is a caring one, then the provision is far more likely to be well-used and therefore likely to bring a reduction of distress to those who participate. It is important to note, though, that counselling needs to be part of an overall strategy of staff care if it is to be fully effective. On its own it may be seen as 'tokenistic' or an attempt to blame staff for feeling stressed as if this were simply a matter of personal weakness or inadequacy (Thompson, 1991).

Employee Assistance Programmes (EAPs)

Many medium and large organisations currently contract workplace counselling to outside consultancy or counselling companies. Research shows that counselling typically brings about reductions in sickness absence, and that after counselling employees feel improved mental and physical well-being, but they show no increase in job satisfaction (Reddy, 1992; Cooper, 1995b). Such schemes do not remove job or organisational sources of stress, which typically requires senior management intervention. Wagenaar and La Forge (1994) comment on the lack of wisdom in an over-reliance on individual-focused therapeutic interventions:

> *In her succinct critique of popular stress manuals, Roskies (1983) cited a ludicrous example of relaxation as panacea. The patient's daughter had just tried to commit suicide when her husband abandoned her; the daughter had also had an accident in which the person in the oncoming car was killed. The patient's husband was dying from coronary heart disease and diabetes. ... Still, despite these monumental problems, the patient continues to function well, enthused the therapist-author of this case, to the point of successfully carrying on her husband's business because, apparently, she "routinely practices her relaxation an hour per day."* (p. 23)

Increased responsibility/opportunities for promotion

It is important in terms of staff care for ability, experience and development to be rewarded by increased responsibility or promotion if appropriate. If staff feel restricted or hemmed in by a 'glass ceiling' this can considerably increase their pressure and decrease their job satisfaction.

Preparation for leaving/retirement

In social welfare we may not be too good at beginnings or at middles, but we are truly awful at endings. Every year thousands of managers, practitioners, volunteers and carers leave social welfare work. Sometimes this leaving is expected (promotion, early retirement, and so on), and sometimes it is unexpected (sickness, death, stress, assault and so on). Expected or unexpected, this leaving is seldom planned or prepared for. Leavers take their stressors into another arena, frequently leaving valuable ways of coping behind.

Exit interviews

Exit interviews are formal mechanisms to prepare for leaving a work situation. They are valuable in that they allow the staff member to de-brief their positive and negative thoughts about the job that they are leaving behind. Not only does this bring some sense of completion, but also it values staff contributions and it allows them to leave behind some of the more stressful aspects of their work experience.

Teamwork

Social welfare work is often undertaken in a team context. That is, staff often work together as part of a collaborative endeavour to achieve shared goals. In terms of the potential for stress, teamwork can be seen as a 'mixed blessing'. Teams can, basically, do one of two things (or do both at different times):

1. Act as a source of a great deal of support, thereby enhancing coping resources and confidence.
2. Provide additional pressures as a result of conflicts, ill-feeling and resentment, thereby increasing the likelihood of stress being experienced.

In view of this two-sided aspect of teamwork it is very important that we do not accept a naive, simplistic and uncritical perspective which sees working in teams as necessarily a good thing. It is certainly true that teamwork can be an excellent way of fending off stress but the reality of the situation is much more complex than this. We therefore need to look quite closely and critically at the question of teamwork so that we do not allow a rose-tinted view to prevent us from seeing some of the pressures and problems that arise in attempts at collaborative working.

One aspect of teamwork that can be particularly difficult and problematic is *multidisciplinary* teamwork. Staff from different disciplines and professional backgrounds can bring different assumptions, values, objectives and patterns of working. Consequently, there is considerable potential for 'sparks to fly' when multidisciplinary teamwork is attempted (see Murphy, 1995, for a discussion of this in relation to child protection). Often such problems can be seen to arise from poor communication or an unwillingness to appreciate other people's points of view or their differing pressures.

There is a danger that a lack of appreciation of other people's pressures can lead to an escalation in terms of conflict or poor team relations. If I am under pressure and I therefore take no account of your pressures, then you may feel undervalued and unsupported and so you may then become less willing to help me with my pressures or understand my point of view. The difficulties then become more entrenched and, once again, up goes the potential for further stress to be experienced.

Good teamwork therefore needs to be premised on an understanding and appreciation of other people's perspectives and the particular demands, challenges and perspectives that they face. Equally, good teamwork needs to be based on a willingness to inform other people of our pressures in order to create an open, responsive and supportive work environment.

While teamwork is clearly a very relevant issue with regard to stress issues, so too is the absence of teamwork. Where individuals work alone without a recognised team or staff group to identify with, there can be problems that arise from isolation and the lack of opportunities to discuss issues, check out ideas, compare notes and generally benefit from the ethos of mutual support that characterises good teamwork. For 'singleton' practitioners who do not have a ready-made team to draw on for support, there is an increased need for networking, for forming links – perhaps both formally and informally – with others who can offer the benefits of team support as and when required. Similarly, established teams may feel it is helpful and appropriate to form links with isolated practitioners with a view to gaining mutual benefit from the possibilities for collaboration that arise.

■ You owe it to yourself

What has clearly emerged so far is that managers (as well as trainers and practitioners) have a lot of responsibilities as far as stress management is concerned. A question this may well raise is: 'Why bother?'. The burden may seem to be so great that it leads to a defeatist attitude in which managers are tempted to leave staff to 'get on with it'. That is, the pressures a manager faces may well leave him or her feeling unable to take on board the pressures faced by staff. Whilst this is an understandable response, it is certainly not a helpful one.

What needs to be recognised is that ignoring staff pressures only leads, sooner or later, to more pressure for the manager. Ignoring stress, or the potential for stress, only serves to store up problems for later – and they have a habit of resurfacing at highly inconvenient moments! It is therefore very much in the manager's interests to tackle pressure and stress proactively, to take control of the situation as far as possible. In short, you owe it to yourself to rise to the challenge of stress.

Stress management training

Training and staff development

So far in this manual we have covered a lot of important points and issues. However, reading about these is only the beginning of a process; it is not enough on its own. To help to bring these issues to life from the printed page, a range of training exercises is presented for you to use or adapt as you see fit.

Working together

The major value of training exercises about stress is that training inevitably involves working together, in partnership, and partnership is a fundamental aspect of good stress management. If stress is not to be a major problem, we need to work together in a spirit of openness and collaboration. These exercises are designed to help you move in that direction.

A journey from one place to another

In recent years a good deal of theory about training and staff development in social work has been discussed. But, for the purposes of this manual, training is simply the vehicle to begin a journey from one place to another. The journey involves developing our understanding about stress in social welfare, so that we can better equip our staff and our organisation to reduce its potency and help with its effects. All of the exercises below are designed to help us with that journey, but they are concerned with different aspects of that journey, some with raising individual and team awareness of stress as an issue, some with an audit of what our stressors are. More are concerned with the control and reduction of stress and the increasing of coping mechanisms. Others are more specifically related to managerial issues and the provision of staff care.

Running the exercises

Each exercise is described in outline, with suggested timings. Please note, however, that timings can vary considerably depending on the nature and size of the group. If you are not experienced in providing training or you don't feel very confident about training issues, it is worth considering collaborating with a colleague and/or drawing on the support of your organisation's training staff. Sometimes we presume that all training will automatically have a beneficial effect on all concerned, but this is not the case – training can have a negative impact on individuals, teams, managers and trainers!

■ Getting the context right

A significant element in delivering positive training is to ensure that the training context is appropriate and positive. There are six significant elements:

1. Training needs
2. Training group (participants)
3. Trainers/facilitators
4. Planning
5. Ground rules
6. Promoting equality/anti-discriminatory practice

1. Training needs

Before attempting to plan any training events, it is essential to find out what your participants need and what they have already done. If training needs are not established, there is a danger that the training will have an inappropriate content, be pitched at an inappropriate level or will duplicate what has already been delivered. This information can come from staff themselves via team meetings or questionnaires, managers who can comment on the needs of the whole team or training departments that may hold information on training needs and previous training that has been run.

2. The training group

It is most important, particularly in training around stress, to be careful about the selection of participants. The wrong selection of participants can lead to the group as a whole not having a productive experience, or even individual members of the group having a very negative experience. When selecting participants it is important to be especially wary of:

- Participants who have been forced to come or really don't want to be there. There is usually a very good reason for this and, if obliged to stay, these participants could seriously sabotage the event or have a very negative experience themselves.
- Having a very uneven group will often lead to participants feeling isolated. If you are the only woman/black participant/manager on the course the potential for isolation and for scapegoating is high.
- Participants who are in the middle of an acute stress episode themselves. Sympathetic managers will often, mistakenly, send very stressed staff on stress management courses, for the best of intentions. Unfortunately, the middle of such an episode is not the best time to learn about stress in a group (referral to a staff counselling scheme at this stage is likely to be much more appropriate – see Practice Focus 2.1).

3. The trainers/facilitators

When deciding who should be responsible for leading the session the following considerations should be borne in mind:

- It is usually not appropriate to have a single trainer. Not only can it be very stressful for the individual, but it ignores the fact that in stress training it is commonplace for one trainer to need to spend time with individual participants. If only one person is training, this can mean that either the whole group grinds to a halt while the individual is cared for, or the group carries on while the individual may 'sink'.
- The trainers do not have to be 'experts' in training or in stress, but it is important that they have an interest in one, or preferably both.
- The trainers do not have to have worked together before – but they do need to have developed an element of understanding and trust around the subject area.

PRACTICE FOCUS 2.1

Helen had been sent on the two-day stress course because she was in a state of acute stress. The trainers had not been aware of her state before the course started, but became worried about her during the initial round. At the first coffee break one trainer spent some time talking to Helen on her own and it became clear that she was too stressed to benefit from the course. It was agreed that she should not continue, but instead was able immediately to go to see the staff care coordinator, who set up a first session with an individual counsellor. Helen returned to do the course six months later after she had completed some very positive individual work.

4. Planning

It is important that the trainers/facilitators spend some time together planning the content and delivery of the training package. We recommend that the content should mirror as closely as possible the training needs that have been established. It is also very important that the style should reflect the needs of the participants and the ability and confidence of the trainers.

5. Ground rules

It is most important, right at the start of the training event, to establish the rules of conduct or 'ground rules' that will govern the behaviour of trainers and participants. These conventions are there to inform people what they are to expect and how they are supposed to behave, but also they are there to keep the participants and the trainers safe. Because stress is a painful and personal issue, the potential for difficulties occurring can be quite high.

There are two different ways of establishing ground rules. The first is to ask the participants what 'rules' they would like to establish and put them on a flip-chart at the front of the room. The second is for the trainers to pre-plan the ground rules and offer them to the group. The former has the advantage of group ownership and contribution, but the disadvantage of time and uncertainty about what you are about to get. The second has the advantage of time and an inclusive list, but the disadvantage of the conventions being imposed on the group to a certain extent. Whichever method is chosen, we would argue that it is important to include the following rules:

- **Communication:** Feel free to give your opinion and to disagree with the opinion of others, without putting them or their opinion down.
- **Commitment:** Be here and work hard for the duration of the course. You are responsible for your own learning.
- **Contribution:** Be aware of how much or how little you are contributing. If you normally say little, try to give more. If you are giving too much, try to give a little less in order to give other people a chance.
- **Confidentiality:** It is a good idea to share the general issues that arise with colleagues outside the training group. However, any individual or personal information about real people will remain within the group. The only limit to this convention is if the trainers believe that there is a serious danger to the course participant or their clients, in which case confidentiality will be broken after informing the participant.
- **Care:** Take care of yourself and the other people in the group. The trainers will agree to 'look after' the whole group to try to ensure a positive experience, but they need you to help care for yourselves.
- **Person-centred:** In this course it is acceptable not to be client- or child-centred, but instead to be person-centred. This course is largely about how you, as a person, are dealing with the professional and personal pressures in your life.
- **Culture:** In our society whole groups of people are devalued and given less status than others. This can be because of their race, gender, age, disability or sexuality. On this course we will attempt not to devalue, by what we say or do, any of these groups of people.

6. Promoting equality/anti-discriminatory practice

The final ground rule in Section 5 above is particularly significant. In so far as a failure to recognise the need to promote equality and challenge discrimination can produce a learning environment which is:

- oppressive for certain groups of people or individuals
- tense, uncomfortable and not conducive to learning
- a potential source of stress in its own right.

We would therefore argue that issues of inequality, discrimination and oppression are taken very seriously and fully integrated into all aspects of the training process.

PRACTICE FOCUS 2.2

Pam was a very experienced practitioner who had attended a number of training courses over the years. She had generally found them useful and enjoyable, although there had sometimes been comments or a turn of events that marred the occasion for her. These usually arose in relation to a lack of sensitivity to the significance of gender and the prevalence of sexist assumptions. In considering whether or not to attend a course on stress she wondered whether the importance of gender would be recognised.

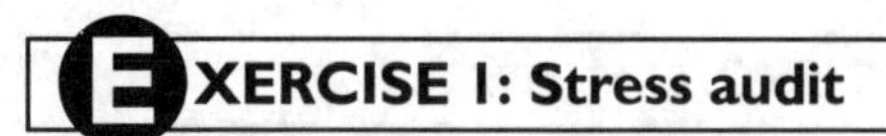

Timing: One and a half to two hours.

Aim: To help staff gain an overview of the stress factors they face.

Structure: An exercise in three parts.

Part One

Introduce the group to the three-dimensional model of stress (OHP 1) and give them the opportunity to ask questions and to seek clarification.

Part Two

Divide the main group up into subgroups (five in each is ideal if this can be managed) and equip them with flipchart paper and pens. Ask each group to brainstorm: stressors, coping methods and sources of support (three columns on the flipchart makes this easier to record).

Part Three

Plenary feedback and discussion. In discussing stressors, try to distinguish between those that can be controlled and those that cannot. For coping methods, try to distinguish between those that are helpful and constructive (for example, talking things over) and those that are destructive, potentially at least (for example, aggression). Consider also a shortfall of coping methods or an over-reliance on one or two 'old favourites'. For support, try to distinguish between formal and informal support and consider how the two can reinforce each other. It is also useful to identify gaps in support systems.

Comment: This exercise is likely to generate considerable discussion and will need to be carefully handled. The plenary discussion needs to be structured and focused. If possible, build in extra time to allow for the exercise 'running over'.

EXERCISE 2: What can I do? What can they do?

Timing: One and a half hours.

Aim: To develop the basis of a stress management 'partnership' by discussing the appropriate allocation of roles and tasks.

Structure: An exercise in three parts.

Part One

A brief introduction to explain the rationale for the exercise and what it consists of (OHP Template 2 can be used).

Part Two

Staff are divided into pairs (or trios, if this makes it easier to divide the group up) and asked to begin filling in the worksheet provided (OHP Template 2 can be photocopied for this purpose). The task is to identify what part the individual ('I') can play in stress management and what part the organisation ('THEY') should play. This should take approximately 45 minutes.

Part Three

Plenary feedback and discussion to highlight the key elements of stress management – both individual and organisational. This will allow strengths and weaknesses to be identified.

EXERCISE 3: Boundaries

Timing: One and a half hours.

Aim: To understand the importance of boundaries of responsibility in stress management.

Structure: An exercise in two parts.

Part One

Using OHP 3, the notion of boundaries of responsibility is introduced. It is explained that some things are the responsibility of the individual (the I circle); some are a shared responsibility (the WE circle); and the responsibility for some things lies beyond our control (the THEY circle).

A key point which needs to be emphasised is the need to recognise, and work within, these boundaries if stress is not to be a problem.

Stress can arise in two ways:

1. If a responsibility within a particular sphere is not fulfilled, then undue pressure may result. For example, if I neglect a responsibility within the 'I' sphere, I may cause problems for myself or for other people.
2. If a person takes responsibility for things beyond his or her control, again undue pressure may result. For example, if budget cuts mean that a certain service or facility cannot be offered, it is important that the individual employee should not shoulder the blame for this.

Part Two

The second part consists of a discussion of the implications of this model. Participants can be asked to give examples of each of the three spheres, and then to consider how they inter-relate. There are (at least) three important points to be drawn out in the discussion:

- Although the three spheres are separate, they are also interlinked – the circles intertwine (as OHP 3 illustrates). That is, we need to understand how individual responsibilities (I) feed into collective responsibility (WE), and how, collectively, staff can influence the organisation (THEY).
- The range of implications for staff (and the organisation) if or when these boundaries are broken or overlooked. That is, what are the costs of ignoring these boundaries?
- Where there is a problem about boundaries, what can be done about this? What strategies could be used?

Comment: This exercise is very compatible with Exercise No 2, and could usefully be used in tandem with it. There is plenty of scope for discussion and the groups should not 'dry up'. However, if this should happen, the group can be divided up into smaller groups to discuss the issues and then feed back to the main group.

EXERCISE 4: Responsibility for/responsibility towards

This exercise is an alternative to Exercise 3 and achieves the same result with a different process.

Timing: One hour.

Aim: To explore the boundaries of individual responsibility and to reduce individual anxiety by examining the appropriateness of beliefs about responsibility.

Structure: A paper and pencil/discussion exercise.

Give the participants a blank piece of paper and ask them to divide it into three. On the top of each section write 'Responsibility *for*', then 'Responsibility *towards*' and finally '*no* responsibility at all'. Ask the participants to write down the major home and work tasks and events that they perform on one section of the sheet or the other.

When the participants have finished, suggest some extra events and tasks, for example: doing appropriate recording; turning up to meetings on time; keeping appointments; treating clients with respect; keeping to deadlines; making sure I provide adequate care for my children; talking to my partner and children (all these should be included in the responsibility *for* section). Then suggest some other events: a client becoming angry; a client committing an assault; a court case that did not go well; a client's death; a planning meeting that felt awkward; not getting on with a colleague or manager; family row; my partner's redundancy and so on (all these should go in the responsibility towards or no responsibility sections).

Discuss with the group the importance of being clear about different kinds of boundary.

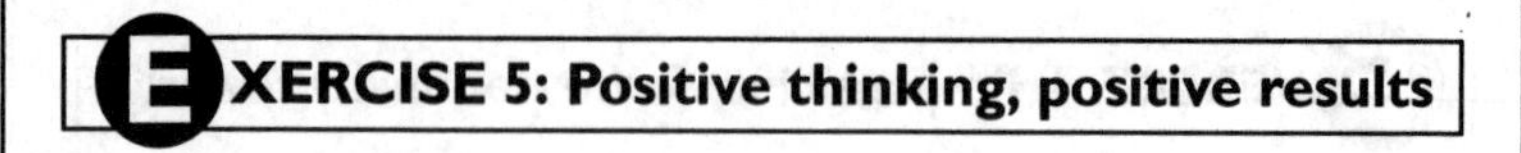

EXERCISE 5: Positive thinking, positive results

Timing: One and a half hours.

Aim: To encourage positive thinking as a safeguard against stress.

Structure: An exercise in three parts.

Part One

A brief introduction to explain and emphasise the importance of positive thinking as a safeguard against stress. OHP 4 can be used to introduce the SWOT analysis – a process by which the Strengths, Weaknesses, Opportunities and Threats of a particular set of circumstances are identified, to give a balanced overview of the situation (rather than focusing specifically on the problems and weaknesses).

Part Two

Divide the group up into small groups and set them the following tasks:

1. Identify a situation which is worrying one or more members of the group.
2. Undertake a SWOT analysis of the situation, paying particular attention to identifying the strengths and the opportunities (the positives) – flipchart paper can be used for this.

Part Three

Plenary feedback and discussion to highlight the key issues and illustrate the difference a positive approach can make.

Comment: This exercise may need 'active tutoring' as many people find it difficult to see beyond the negatives of a situation, particularly professional 'problem-solvers' such as social welfare workers. Participants may need a lot of encouragement to think creatively and recognise strengths and opportunities.

EXERCISE 6: Team building

Timing: Variable, from one hour to a whole day.

Aim: To encourage a supportive team atmosphere.

Structure: Variable.

Team-building exercises are many and varied and can be tailor-made to suit the circumstances of the team or staff group. Broadly, there are two types of exercise, person-centred and task-centred. One example of each type is given here.

Person-centred

A useful get-to-know-you exercise can be undertaken by asking each staff member to complete a profile form (see next page) in advance of the team-building session. Each person in turn shares the comments on their form with one or more colleagues, and this allows common areas of interest or background to be identified. It can also be a fun exercise and therefore a good icebreaker.

The centre of the form is for personal details (name, age and so on). The top section is a summary of experience; the bottom section a summary of hopes, fears and aspirations. The left-hand side is for recording positives (the things I like about my job/situation) and the right-hand side for the negatives (the things I don't like).

Task-centred

Building an 'objectives tree' is a helpful way of getting a sense of common purpose and shared commitment. A flipchart or board is used to draw a map or tree of the team's objectives. This is done step by step from top to bottom; from the general, overall aim (for example, high quality services) to more specific objectives (see OHP 5). The group members are asked to agree their overall aim and then break this down into its component parts, and then break each of these down into their component parts, and so on. OHP 5 gives one framework for this but the number of boxes at each level will vary according to the circumstances.

Comment: How these exercises are to be used (for example, timescale) needs careful planning in order to avoid drift and vagueness. However, such planning can be undertaken by the team so that a spirit of sharing and partnership is encouraged right from the start.

Profile form

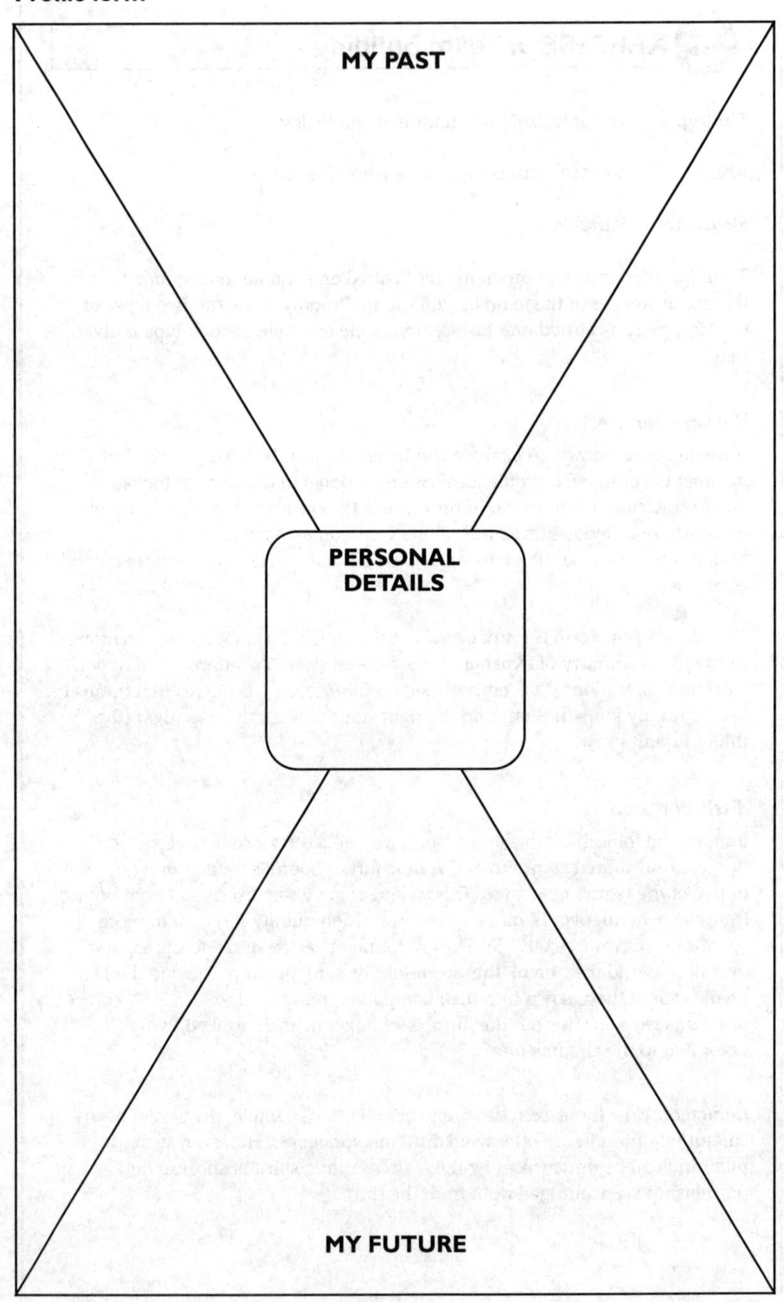

EXERCISE 7: Quality circles

Timing: Half a day per month.

Aim: Improving quality and problem-solving.

Structure: Unstructured or semi-structured, depending on the circumstances.

A quality circle is a group of people who meet on a regular basis (half a day per month is a common pattern) to discuss particular problems or issues. The group takes ownership for solving particular problems or tackling issues of quality assurance (hence the name). Quality Assurance is a broader approach, defined by ISO 8402 as:

> *'All those planned and systematic actions necessary to provide adequate confidence that a product or service will satisfy given requirements for quality.'* (see Bone, 1994).

This approach has the advantages of:

- encouraging a shared, mutually supportive approach;
- empowerment – encouraging professional autonomy;
- stimulating creativity and imaginative approaches to problem-solving;
- creating opportunities for job satisfaction.

Quality circles can play a significant role in stress management in two ways, general (indirect) and specific (direct). The general way is through the use of quality circles to tackle a whole range of work-based problems, thereby providing possibilities for reducing pressures by changing practices, enhancing coping methods and optimising support. More specifically, a quality circle could be established to address issues of stress directly and explicitly.

Comment: If quality circles are to work to full effect, they need to be planned carefully in terms of group composition, timescale, problems to be addressed, priorities and so on. Time spent in planning will be a worthwhile investment, whereas rushing into an ill-conceived quality circle could cause more problems than it solves. Most quality circles work best when a 'facilitator' has been chosen to help the group process get started and to act as a 'go-between' with senior management. Quality circles are most effective when the culture of the organisation is such that:

- power is readily devolved to staff groups;
- the circle sessions do not become whinging sessions;
- the work of the circle is seen as important;
- the circle is given adequate time (say six months) to develop.

EXERCISE 8: Atmospheres

Timing: One and a half hours.

Aim: To promote a positive working environment.

Structure: An exercise in two parts.

Part One

Divide the group into small groups and set them the task of identifying what factors contribute to a positive working environment (for example, good channels of communication) and what factors detract from it. These can be recorded on sheets of flipchart paper. Groups may want to draw a line down the centre of the sheet and list positives on one side of the line and negatives on the other.

Part Two

Plenary feedback and discussion. It is helpful if discussion focuses on strategies for minimising the negatives and maximising the positives.

EXERCISE 9: What, how, when

Timing: One and a half hours.

Aim: To increase confidence through developing a systematic approach to one's work tasks.

Structure: An exercise in four parts.

Part One

The 'What How When' framework is introduced as a means of developing a clear systematic approach to work tasks which can:

- give a clearer sense of purpose and direction;
- increase control over work pressures;
- reduce anxiety and confusion;
- boost confidence.

OHP 6 can be used to explain the framework, setting out the three key questions which need to be addressed in planning or reviewing particular tasks or projects.

Part Two

Each participant is asked to write down brief details of a work task or situation which is currently causing concern. This should only take a few minutes.

Part Three

Divide the main group into trios and set them the task of 'reframing' each participant's identified concern in terms of the three key questions (Thompson, 1996):

- What are you trying to achieve?
- How are you going to achieve it?
- How will you know when you have achieved it?

Part Four

Plenary feedback and discussion. This can usefully focus on the benefits of using the 'What How When' framework as a means of managing work pressures and ways and means of dealing with any obstacles to using this approach.

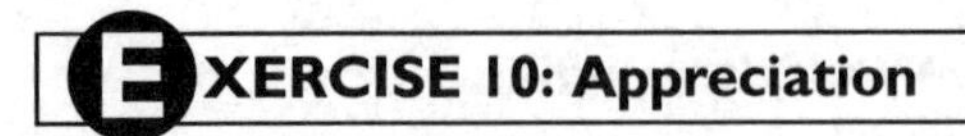

EXERCISE 10: Appreciation

Timing: Thirty minutes maximum.

Aim: To encourage a supportive atmosphere in which staff feel valued and appreciated.

Structure: A brief and simple exercise which can be used to end a staff meeting, team-building day or training course.

You will need to provide a medium sized envelope for each participant. Write a participant's name on each envelope. Lay out the envelopes on a table or other work surface so that participants can have easy access to them. Then, set the task of every person making a positive comment about each of the other individual group members on a slip of paper and 'posting' them within the appropriate envelopes. Comments can include personal characteristics, skills, knowledge, attitudes and so on. They may be specifically work-related or more general. At the end of this, give each person their own named envelope so that they can read the positive comments made about them.

Comment: It is not advisable to use this exercise with a group where there are significant underlying tensions or conflicts – it could backfire!

Be sure to emphasise very strongly to the group that only positive comments are allowed. Critical or ambiguous comments are to be strictly forbidden, even in jest – what is intended as a joke can appear very different to the person receiving an anonymous comment in an envelope.

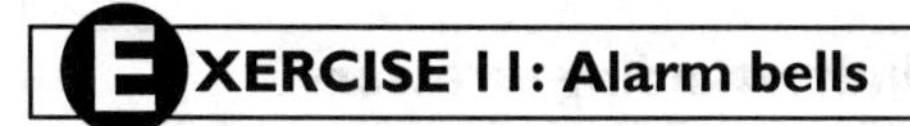

EXERCISE 11: Alarm bells

Timing: One and a half hours.

Aim: To identify signs of stress in others.

Structure: An exercise in three parts.

Part One

The exercise begins with the facilitator emphasising the crucial role of mutual support in stress management. Part of this mutual support is the ability to recognise signs of stress in others – and this is the basis of the exercise, to explore ways and means of recognising stress.

Part Two

Divide the main group into sub-groups and equip each group with a sheet of flipchart paper and a marker pen. Set the group the task of listing possible signs of stress that can be recognised in colleagues (and that they can recognise in us).

Part Three

Plenary feedback and discussion. This can usefully focus on three key issues:

- Different people show their stress in different ways – 'know your staff' is an important maxim.
- Despite these differences, there are also commonalities, common themes that we would do well to remember and be sensitive to.
- Whilst it is important to recognise stress in others, we should not lose sight of the importance of recognising stress in ourselves.

EXERCISE 12: Understanding stressors

Timing: 45 minutes.

Aim: To allow the exploration of individual stressors.

Structure: An exercise in two parts.

Part One
Ask the participants to fill out as much of the personal record form as they choose (see next page).

Part Two
Ask the participants to work in pairs and discuss the work aspects of the record form.

Part Three
Feedback and discussion based on asking the participants to identify the stressors that they held in common.

Comment: This exercise allows time for individual participants to explore their stressors and to discuss them with one other person. The feedback serves to emphasise the fact that, although stressors are unique to the individual, there are many that we can hold in common with others.

Stressors exercise record form

	AT WORK	AT HOME
What are my stressors?		
How do they affect me?		
What could I do to reduce them?		
What stops me doing this?		

EXERCISE 13: Avoid, minimise, control

Timing: 30 minutes.

Aim: To allow individuals to think about changing their own stressors.

Structure: An exercise in three parts.

Part One

Part One begins with a brief spoken input: explain to participants that individual stressors, even though we are used to dealing with them, are often able to be reduced or controlled. This might be via:

a) *Avoidance* – the best way to reduce the power of a stressor is to consciously avoid it.
b) *Minimising* – Where we cannot avoid a stressor completely, we can often reduce its power to pressurise us.
c) *Control* – Where the stressor is not avoidable or open to minimisation, we can seek to control it by having the power to decide where, when, how and with whom we face it.

Part Two

If participants do not already have a list of stressors, ask them to use the personal record from Exercise 12 to identify them. Ask the participants to look again at their list of stressors. Ask them to put a mark against those stressors which they might be able to reduce.

Part Three

Ask the participants to get into pairs, choose one stressor each and discuss how they might change it.

EXERCISE 14: Great expectations

Timing: 15 minutes.

Aim: To reduce stress and anxiety by reducing the participants' expectations of themselves.

Structure: An exercise in two parts.

This exercise is a guided fantasy, and therefore can be extended beyond 15 minutes if you wish.

Part One

Begin with any form of relaxation exercise to help the group to unwind and relax. As a minimum, it might include an invitation to:

- sit down.
- close your eyes or fix them on one part of the floor.
- take a series of deep breaths in, followed by the same number of long, slow breaths out.
- take notice of your body – where is it tense? Imagine that tenseness gradually seeping out of your feet into the floor.
- pay attention to your mind – what are the worries that are racing round within it? Take them one by one, wrap them up in brown paper and put them into a big brown cardboard box, to be opened later.

When the group feel ready (if they are enjoying the relaxation let them have a few minutes to enjoy the sensation of relaxation) move on to Part Two.

Part Two

Part Two is also a guided fantasy. Imagine that you are in a warm, quiet room, sitting on a comfortable chair facing a large blackboard. Right at the top of the blackboard is your name with *'Performance Expectations'* next to it. Under this there is a series of charts with *'Partner'*, *'Team Member'*, *'Parent'*, *'Friend'* and so on. Each chart shows expectations of 100 percent in each area. Take the board duster that you find underneath your chair, go up to the board and rub out 15 per cent of the 'expectations' that you find on each chart – each now reads 85 per cent instead of 100 per cent. Go back to your chair, look at the new charts. You may feel a pleasant sense of relief and satisfaction.

EXERCISE 15: Staff care – the why and the whether

Timing: 40 minutes.

Aim: To allow participants to explore the usefulness of staff care.

Structure: This exercise can be run as a straightforward 'brainstorming' or as a debate between two opposing camps.

Brainstorm

Divide the group into sub-groups of between three and six and ask each group to brainstorm one of the following:

- What use is staff care to modern social welfare organisations?
- At what points in their working life might staff benefit from staff care?
- What might prevent staff from using a staff counselling service?

Allow 15 minutes for the brainstorming, then 25 minutes for feedback and discussion. The discussion can be used to highlight:

- The crossover between good staff care and positive management practice.
- The relatively inexpensive nature of positive staff care.
- The 'cradle to grave' need for staff care.
- The 'blocks' to staff counselling – for example, a macho organisational culture, the 'We're too busy syndrome', 'Others may need it but we don't' and so on. (See Part 1: '25 ways to stress out your staff' and 'Myths about stress'.)

Debate

Divide the group in two, with a convenor for each group. Group One should prepare their arguments to support the proposition that: 'Positive staff care is an asset to staff and to the agency', while Group Two prepare their arguments to support the view that: 'The agency's only role is to pay the staff's wages, not to mollycoddle them'. Position the two on either side of the room, invent an imaginary microphone and run the debate.

■ Conclusion

The training exercises presented here are, of course, not a guaranteed recipe for success. They should, however, provide a range of opportunities for exploring important issues relating to stress in a relatively safe and supportive environment. The exercises are not fixed and rigid insofar as they may be adapted in a number of ways and used quite flexibly. However, we would advise great caution in amending the exercises without first thinking through the implications of any changes made. For example, shortening an exercise may not leave enough time to draw out the learning points, while extending an exercise may mean that its impact is lost. As with all high quality training, careful planning and anticipation of potential problems must feature significantly.

One final point that we wish to emphasise is that we should beware of seeing training in too narrow a context. We prefer to see it as part of a broader process of staff development that should link in with supervision, appraisal and so on (Thompson and Bates, 1995). If training is to be effective beyond the narrow confines of the training room, it needs to be part of this broader whole, part of a human resources management approach based on principles of staff care.

Conclusion

Recognising stress 3

Are you on top of your job – or is your job on top of you?

Part 3 draws from recent research on stress and distress at work, organising the material in terms of answers to four questions:

1. How satisfying is your job at present?
2. What are the features of jobs which bring work-related stress?
3. What are the common signs of stress and distress in individuals?
4. How can we cope? How can we get support?

We shall address each of these in turn.

■ 1. How satisfying is your job at present?

Work is very important to people. It provides a major part of a person's sense of identity and status and is an important source of satisfaction and self-esteem, but it may also be a source of dissatisfaction. If your job is not satisfying it may be, or become, a source of stress. If you are satisfied with your job at present you would probably agree with the following statements:

- Most days I am enthusiastic about my job;
- I find real enjoyment in my job;
- I like my job better than the average person does;
- I am seldom bored with my job;
- I would not consider leaving to take another job,

and if you are dissatisfied you would probably disagree with them.

There are a number of features of jobs and of people which render employees more likely to report their job as satisfying. Firstly, the type of job you do makes a difference: routine and repetitive jobs are rated as less satisfying, compared to jobs which are creative, require you to keep on learning new things, require a high level of skill and let you use your skills and abilities. That is, jobs which make demands on you, but demands which you can meet, are satisfying and raise your self-esteem.

While nobody would object to being paid lots of money for doing their job, it turns out to be more important for job satisfaction that the distribution of resources within the workplace is seen as fair. Employees, especially in the public sector, generally understand only too well the restrictions on the 'amount of cake' available, but can rate themselves as satisfied if they feel they are getting 'a fair slice' compared to others in the workplace, given their qualifications, experience, responsibilities and effort.

The type of person you are also makes a difference. In general, people who are highly involved with their work, who feel that their work is central to their lives, who live to work rather than work to live are likely to report higher job satisfaction.

And finally, people who are characteristically positive, happy and optimistic, who see their daily life as involving a series of uplifts rather than an unending round of hassles, are more likely to report satisfaction with their current job.

2. What are the features of jobs which lead to stress?

A number of features of jobs have been identified as likely to contribute to feelings of stress at work. These may be classified in several ways. Here we divide them into the *content* of the job – what you have to do, where and when – and the *context* of the job – the organisational setting in which you have to do it.

The presence and intensity of each of these stressors will vary from job to job and from time to time. In addition, the extent to which each stressor is experienced as stressful will vary from person to person.

Content features

Probably the most common source of stress is excessive load or pace of work. Being required to do too much too quickly is stressful, especially if, over a period of time, the load and the pace do not let up. Being required to do too much in the time you have available is a clear case of the demands of the job exceeding your personal mental and physical resources. Demands and resources are no longer in balance, you no longer feel in control. Unremitting demands are likely to make an employee first anxious, next irritable, then depressed, and finally exhausted.

If you are under workload stress you are likely to agree with the following statements:

- I have to work very hard just to keep up with my work;
- I do not have enough time to complete my work;
- I am asked to do things without adequate materials and resources to accomplish them;
- I am asked to do things beyond my training and capability.

All this holds true whether the job involves working on a production line, dealing face to face with customers or clients, or handling paperwork.

The physical setting of the workplace can also contribute to stress. If it is unhealthy, noisy, over-lit or under-lit, if it is badly designed for its function, making the job harder to perform, all these can contribute to the stress experienced by employees. Shift work and long or unpredictable or unsocial hours can all contribute to stressing employees by making demands of them when they are at a low ebb, and thus less well placed to respond. When people are tired they make more mistakes. Mistakes, at the very least, take up more time in putting them right; at worst they may be devastating to the success of the enterprise, or even fatal.

Context features

Uncertainty

Uncertainty is always a source of anxiety, and too much uncertainty is a source of stress. In the recent economic climate uncertainty about whether you will have the same job or, indeed, whether you will have a job at all come next year can be a constant stressor, reducing the enthusiasm, morale, performance and efficiency of employees. Even when your job is secure, uncertainty about what you should be doing and about how well you are doing it adds to the stress. If you are under the stress of uncertainty (often called role ambiguity) you will be likely to disagree with the following statements:

- I receive a clear explanation of what I have to do;
- I know exactly what is expected of me;
- I know what my responsibilities are;
- I am told how well I am doing my job.

Information ('What's happening?'); clear instructions ('What should I do next, and when by?'); and feedback ('How am I doing?') reduce uncertainty, in all walks of life.

Conflicting job demands
Conflicting job demands may arise in two ways. Firstly, where the chain of command has you reporting to two or more superiors who seem to have different ideas about what you should be doing and/or how you should be doing it. If this is the case you are likely to agree with statements such as:

- I receive incompatible requests from colleagues or superiors;
- Jobs I do are accepted by one person but not by another.

Secondly, conflict arises if you are required to do things or do things in certain ways that clash with deeply held values. Here you would likely agree with the statements:

- I am asked to do work that doesn't fit with my values;
- I have to do things that I think should be done differently.

Decision making
In general, stress is about feelings of not being in control of things or not having things under control. In the extreme this can lead to feelings of helplessness and hopelessness. This can be related to decision-making in the workplace in three different ways:

- First, autonomy refers to the amount of freedom employees have to act independently and make their own decisions. Employees who feel they are trusted to act on their own, without constant reference to superiors, normally repay this trust, act responsibly, and derive satisfaction from meeting this challenge, especially when the boundaries between what they can and cannot do on their own initiative are clear and agreed. Employees who feel they are denied proper autonomy find this a source of stress.
- Second, employees who are denied participation in the decision-making of their work unit, team, department or organisation find this stressful. Part of the reason for this is likely to be the uncertainty mentioned earlier which engenders anxiety.
- Third, responsibility at work is a source of stress. Responsibility may be for things such as budgets, records, equipment or buildings, or it may be for people. Having high responsibility for people is especially stressful, particularly when the economic climate makes for conflict between personnel costs and attending to the welfare and well-being of others.

Personal development
If there is little or no opportunity for personal development in the job, this may be a source of stress. On the one hand, personal development may involve the chance through job redesign, training and development to improve skills, abilities and knowledge, increasing one's sense of mastery and providing more variety and challenge in the job. On the other hand, personal development may involve opportunities for internal promotion and advancement, bringing both change and satisfaction. Both types of development are likely to enhance an employee's self-esteem. If an employee sees no prospect of either, they are likely to feel stuck, with no chance of escape, and to feel burdened and undervalued.

Similarly, the state of the job market makes a difference. If there are no other jobs available so that someone could not easily find another job as good as or better than the one they have, then they are likely to feel they have no alternative but to stick with the present job.

Making a contribution
It is important that an employee feels that their work makes a significant contribution, however small, to the overall success of the organisation. If you don't get this feeling from your job, you are likely to agree with the following statements:

- Sometimes I'm not sure I understand the purpose of what I'm doing;
- I don't understand how my work fits in with the work of others here;
- I often wonder what the importance of my job really is;
- I often feel that my work counts for very little around here.

Life changes
Finally, stress at work may be affected by major life changes elsewhere. Some of the largest changes here would be the death of a spouse or close family member, separation from a partner, marriage and pregnancy, major illness in the family, substantial financial problems. All of these are likely to take one's attention away from work, and employees dealing with one of these changes need special care. More routinely, work performance may be affected if there are conflicting demands between work and home, or if the sum of work and home stressors exceeds the person's capacity to cope.

What can and cannot be changed
Another important way of classifying job stressors is the division between those which are remediable – those you or your boss or your organisation can do something about – and those that are irremediable – those that are just outside the control of anyone in your organisation. But in both cases somebody should do something.

Some aspects of a person's job load may be changed – through job redesign, through the intervention of supervision to make loads more equitable, or through the provision of training and/or development opportunities to staff.

Some aspects, though, may be wholly intrinsic to the job – for example, in social welfare, the nature of the clientele – or not susceptible to change within the organisation – for example, statutory responsibilities that the organisation is charged to fulfil. For these, the employer must support ways of enabling staff to cope with the 'givens' of the job.

■ 3. What are the common signs of stress and distress in individuals?

How can you tell when you are stressed?

A recent study of teachers reported that: "Some talked of tiredness, irritability and depression, of sleeping badly, increased drinking, occasional crying in the staff room, and a sense of guilt that they were neglecting their families." Thus there is a whole range of types of 'symptom' which have been characterised in many ways:

As *mental symptoms*, for example:

- an inability to concentrate;
- difficulty in making decisions;
- memory lapses and errors.

As *emotional symptoms*, for example:

- irritability or outbursts of anger;
- feelings of anxiety or insecurity;
- moodiness;
- fear of criticism.

As *physical symptoms*, for example:

- headaches;
- tearfulness;
- tenseness;
- feeling tired and run down;
- restlessness;
- inability to relax.

And as *behavioural symptoms*, for example:

- increased smoking or alcohol consumption;
- changes in eating and sleeping patterns;
- reckless driving;
- absenteeism or workaholism.

The only good thing about stress is that, if you are really under stress, your body wastes no time in letting you know, and letting you know in a whole variety of ways.

Use the Stress Checklist on the next page to assess your own level of stress.

Stress Checklist

To assess your current level of stress, indicate how often you have been troubled by the following 'symptoms' within the last three months:

0 = Never or rarely
1 = Occasionally
2 = Frequently
3 = Always or nearly always

Constant irritability with people	0	1	2	3
Difficulty in making decisions	0	1	2	3
Loss of sense of humour	0	1	2	3
Suppressed anger	0	1	2	3
Difficulty concentrating	0	1	2	3
Unable to finish one task before rushing into another	0	1	2	3
Feeling yourself the target of other people's animosity	0	1	2	3
Feeling unable to cope	0	1	2	3
Wanting to cry at the smallest problem	0	1	2	3
Lack of interest in doing things at home	0	1	2	3
Feeling tired after an early night	0	1	2	3
Constant tiredness	0	1	2	3
Lack of appetite	0	1	2	3
Craving for food when under pressure	0	1	2	3
Indigestion or heartburn	0	1	2	3
Constipation or diarrhoea	0	1	2	3
Unable to get off to sleep	0	1	2	3
Sweating for no good reason	0	1	2	3
Nervous twitches, nail biting, etc.	0	1	2	3
Headaches	0	1	2	3
Cramps and muscle spasms	0	1	2	3
Nausea	0	1	2	3
Breathlessness without exertion	0	1	2	3
Fainting spells	0	1	2	3
Impotent or disinterested	0	1	2	3
Eczema	0	1	2	3

Scoring: It is not the total score which is important, but the number of 'symptoms' on which you score 2 or 3. If you are showing more than 6 'symptoms' with scores of 2 or 3, then you may have a current stress problem.

[adapted from Cartwright and Cooper, 1994, p. 13]

■ 4. How can we cope? How can we get support?

There are two questions to answer in dealing with stress. The first is: Am I being stressed here? The second is: Can I cope? If the answer to the first question is "Yes", and the answer to the second is "No", you're under stress.

The most direct and efficacious way of dealing with stress is either to remove the stressor from the person or to remove the person from the stressor (so that, either way, they are no longer *under* stress). That is, permanent solutions should deal with the underlying root cause, rather than the surface manifestations or 'symptoms'. However, in the short term, if the problems are excessive, then they may well need specialist intervention (for example, counselling).

There is a large range of coping techniques and strategies that people can and do use for reducing stress at work. Some of these focus on dealing with the problem or problems, some focus on dealing with emotional reactions to the problem. Both attempt to restore a sense of control: for the former, control over the situation and, for the latter, control over one's feelings.

Short-term strtegies
Almost all coping strategies offer some short-term gain, but not all are beneficial in the longer term. Examples of the latter are:

- absenteeism or taking unofficial time off work;
- turning to drink;
- excessive or 'comfort' eating;
- denying that there is a problem.

These are often called 'avoidance' strategies. They give some breathing space and serve to reduce the immediate distress by distancing oneself from the problem, but the problem still remains.

More positive strategies
There are more positive and, in the long-run, more helpful ways of coping.

For dealing with the pressures of work, better anticipation and planning, more efficient time management and prioritising of work tasks should help.

For dealing with feelings of anxiety and uncertainty, careful monitoring of one's reactions (by, for example, keeping a 'stress diary' recording pressures, incidents and the feelings engendered) can help with getting straight what, exactly, is the problem. There are a large number of breathing and muscle relaxation techniques which serve to reduce the intensity of the distress felt. Keeping physically healthy through exercise and a balanced diet helps maintain a store of energy for addressing the problem. Successful completion of home and leisure tasks and activities provides healthy distraction from the pressures at work and helps restore a sense of competence, control and self-esteem.

Support from others can enhance one's personal resources for dealing both with the stressors and with the distress. At work, official support in the form of sympathetic supervision, re-scheduling of workloads, training and guidance in new skills all improve the ability to cope with the demands of the job. Unofficial support from colleagues and workmates reduces the sense of isolation and helplessness and improves the ability to handle upsetting feelings. Support outside work from partners and friends can be vital in putting things in perspective and helping to restore one's balance by re-assessing the demands you face and your capacity to cope with them.

In brief – job satisfaction

People are more likely to be satisfied with their jobs when they feel that:

- work is not mentally and physically too tiring;
- working conditions are compatible with physical needs and facilitate meeting work goals;
- rewards (whether pay, resources, recognition or feedback) are seen as fair and just;
- others in the workplace (executives, superiors, colleagues, subordinates) help rather than hinder meeting the aims and values of the organisation;
- work represents a mental challenge they can handle which leads to involvement and interest in the job and gives them a strong sense of accomplishment and self-esteem.

In brief – job stressors

People are more likely to feel stressed at work to the extent that they answer "NO" to the following questions:

- Do you have variety in your job?
- Do you feel you are rewarded fairly for the work you do?
- Do you have enough time and resources to do your job?
- Are you clear exactly what is expected of you in your job?
- Is your job free from conflicting demands from various others?
- Are you free from having to do things you wish you didn't have to do?
- Do you have freedom to make decisions about the way you do your job?
- Are you involved in or informed about important decisions at work?
- Do you have responsibility for other people or responsibilities which can affect the welfare of others?
- Do you have opportunity to change the job you do at work?
- Could you get a better job elsewhere?
- Are you clear how your job contributes to the work of the organisation?
- Do you have a supportive supervisor?
- Could you confide in the people you work closely with?
- When work gets tough do you get plenty of support at home?

In brief – dealing with stress

Typically there are five stages that individuals go through in addressing their own stress – and that managers go through in addressing the stresses of those they are responsible for:

- Stage 1 – has heard of stress;
- Stage 2 – believes (only) others are susceptible to stress;
- Stage 3 – acknowledges own susceptibility to stress;
- Stage 4 – decides to take action about stress;
- Stage 5 – takes precautions against stress.

The job of management is to ensure the whole organisation functions at Stage 5.

The material in this section is drawn from the following sources:

Agha, A.O., Mueller, C.W. and Price, J.L. (1993) 'Determinants of Employee Job Satisfaction: An Empirical Test of a Causal Model'. *Human Relations,* 46(8), 1007-1027.

Cartwright, S. and Cooper, C. (1994) *No Hassle! Taking the Stress out of Work,* London, Century Business Books, Random House.

Keita, G.P. and Sauter, S.L. (eds) (1992) *Work and Well-Being: An Agenda for the 1990s,* Washington, DC, American Psychological Association.

Patel, C. (1989) *The Complete Guide to Stress Management,* London, Macdonald.

Ross, R.R. and Altmaier, E.M. (1994) *Intervention in Occupational Stress,* London, Sage.

Weinstein, N. (1988) 'The precaution adoption process', *Health Psychology,* 7, 55-386.

Tackling pressure and stress 4

What can *I* do?

This part of the manual is meant for the individual member of staff and can be used as the basis of a handout or explanatory leaflet for whole team training or for specific stress workshops, training courses or other staff development activities. The content of this section is divided into five parts. These are:

1. Acknowledging stress
2. Understanding stress
3. Recognising stress
4. Dealing with stress
5. Protecting yourself from stress

The stress challenge – the individual's responsibility

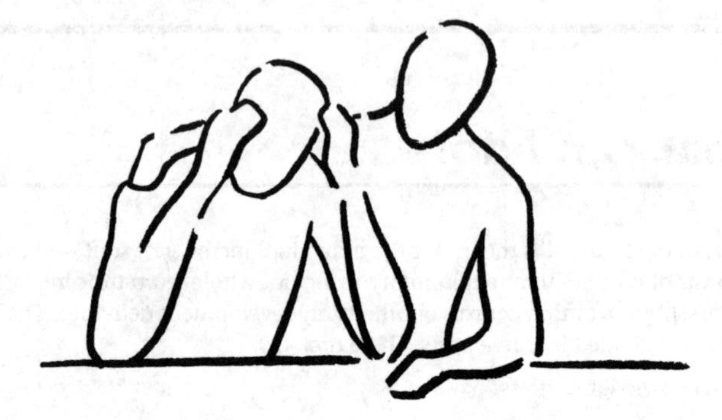

Stress is not just the business of agency, manager or team. Stress is the business of the individual. The stress challenge is for everyone to meet.

■ 1. Acknowledging stress

The first step in meeting the stress challenge is to acknowledge that stress exists and that it hurts and costs us dearly.

- Stress is not a mythical being like Santa Claus or the 'bogeyman'. It does not care if we believe in it or not – it exists independently of this belief. Stress can hurt you even when you don't believe in it.
- Ignoring stress, being too macho, too tough, too professional or too busy to be stressed does protect us – stress gets through anyway.
- Stress will affect all of us at some point in our working lives. It is not a sign of weakness – it is a sign of our humanity. This is why we all have a vested interest in combating stress.
- Unrelieved stress hurts – it debilitates us physically, mentally and emotionally. It can seriously damage our sense of happiness and our self-esteem.
- Stress also hurts those around us, those with whom we have relationships, from our clients and colleagues at work, to our families, partners, children and friends at home. Stress not only exhausts us, it cuts us off and isolates us from those who would normally help.

2. Understanding stress

Pressure and stress

Everyone experiences pressure in their daily lives. This is not necessarily a bad thing. The right amount of pressure can help you to achieve a great deal. Like a following wind, it can encourage you to get where you intend to go.

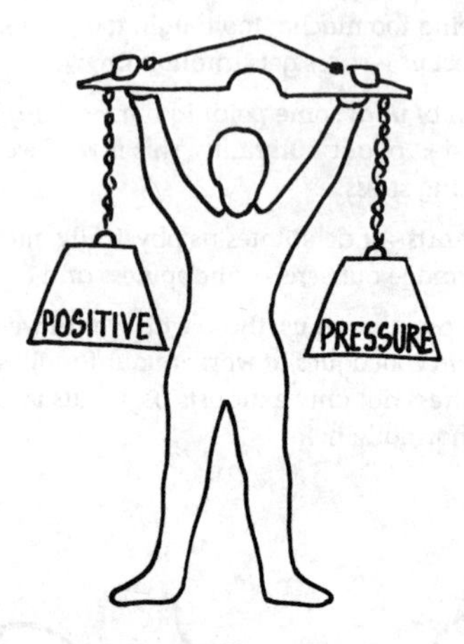

However, a point is reached where pressure outstrips your ability to cope. After this point pressure becomes stress. The more pressure increases, the more stress builds up in the individual, and the more that individual is in danger of becoming exhausted.

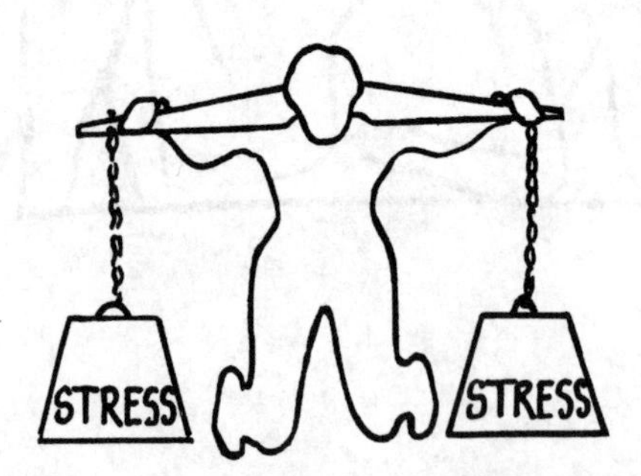

It is the turning of such pressure into 'stressors' that is the crucial negative factor in terms of stress. The more these stressors build up, the more vulnerable we are to debilitation.

However, there are two useful factors that help us to keep pressures from becoming stressors. These are *coping resources*; anything that we do to look after ourselves – and *support systems*; anything that our agency/team/family do to 'look after' us. As long as these two helpful factors can`be kept in balance with the pressures that we face, then our pressures do not become stressors. But when our pressures overload our ability to care for ourselves, stress is the result. This is called the stress balance.

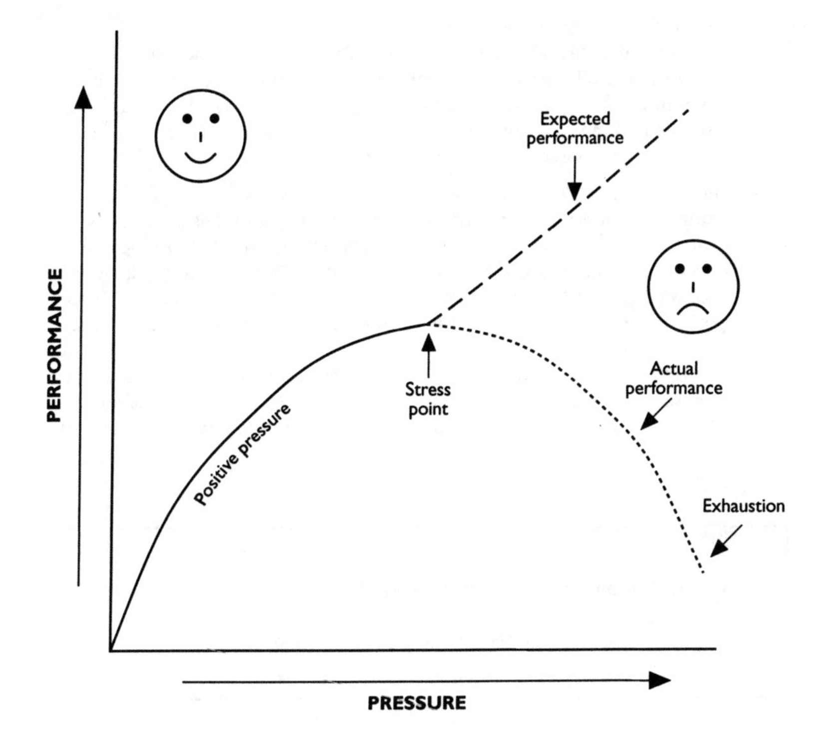

Sources of stress

There are three main sources of pressure and stressors in our lives. These are:

- *The stressors that come from our personal life*
  These include stressors from our relationships with partners, children, family, friends and neighbours. They can include financial or housing difficulties. Anxiety, confusion and uncertainty about our home situation can also add to our stress.
- *The stressors that come from work*
  These can include difficult relationships with managers, colleagues or clients; too much or too little work; harassment, abuse or bullying; a poor working environment or the nature of the work itself – particularly when your work has a high emotional content. Anxiety, worry, confusion and uncertainty about your work situation can also cause stress.
- *Single, serious incidents that occur in your life*
  During the course of your home and work life, single serious incidents will always occur. These can include undesirable changes that might include bereavement, separation or work reorganisation. They also might include positive incidents like promotion or the birth of a child. All these incidents, positive or negative, can seriously intrude upon your life and cause stress.

Unfortunately there are no 'barriers' to these different sources of stress – the stress from all these areas is cumulative. Therefore your total stress is the sum of your general home stress plus your general work stress plus your serious single incident stress.

General Home Stress + General Work Stress + Single Incident Stress

= TOTAL PERSONAL STRESS

3. Recognising stress

Although stress is felt individually, the signs of stress are often held in common. Therefore if you become good at recognising stress in yourself, there is a good chance that you will be able to recognise it in others.

You can find out how stressed you are by listening to what your body is telling you. Stress always shows itself, but sometimes it builds up so slowly that we do not notice the changes involved. We must stop, take time and work out how we are. Are we under appropriate pressure, beginning to show the effects of stress, or are we in danger of serious debilitation?

This is called taking our stress temperature.

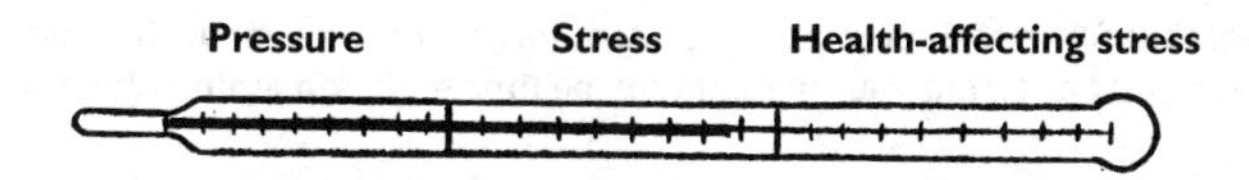

Do you find that you have:

Physical indicators

Are you

- tired/exhausted?
- over/under-sleeping?
- over/under-eating?
- over-consuming substances?
- heart racing?
- tense/unable to relax?
- over-sweating?

Emotional indicators

Are you

- excessively moody?
- dispirited?
- anxious?
- worried?
- feeling helpless?
- weepy?
- constantly sad?
- afraid?
- irritable?
- angry?

Behavioural indicators

Are you

- angry and aggressive?
- hiding away/avoiding people?
- de-motivated?
- muddled in your thinking?
- blaming others?
- taking time off work?
- drinking/eating/smoking too much?
- finding everything and everyone a burden?

If you feel that some of these signs are present for you or a colleague, it could be that stress is seriously affecting your lives. It may be time to do something about it.

■ 4. Dealing with stress

For yourself

If your life is becoming seriously affected by stress, there are steps that you can take to help yourself.

Reducing your stressors

One of the most effective ways of combating stress is to attempt to reduce your stressors both at home and at work:

Avoidance

The quickest and simplest way to reduce our stressors is to try to keep ourselves out of their power by avoiding facing them. So, if a certain work routine always leaves us exhausted – find another way of getting that job done. Or, if getting the children out in the morning is particularly stressful, give yourself an extra half an hour to cut out the rush. Although this is the most logical technique for dealing with stressors, it is not as easy as it sounds. We become accustomed to our stressors and find it hard to avoid them. Changing habits, even those that do us harm, is not an easy task.

Minimise and control

If we cannot avoid the stressor, the next best thing to do is to reduce the power of the stressor and establish some kind of power over it. So, if we know that an interview is likely to lead to violence or confrontation, we can minimise it by getting assistance, doing the piece of work in your territory, on your terms, getting the environment right, making a contingency plan and deciding where your boundaries lie (when do we close the interview/get out/call the police and so on). Many of our general home and work stressors are not completely avoidable, but we can minimise and exert more control over almost all of them.

Change your perception

Another very effective way of reducing stressors is to change how we perceive them. This cognitive technique is based on the presumption that what puts us under pressure is not necessarily the pressure itself but what we make of it and how we see it. This is particularly relevant to stressors that cause us considerable apprehension and anxiety. Rational Emotive Therapy suggests that such anxieties are frequently made up of irrational as well as rational belief systems – if we can remove the irrational beliefs this will considerably reduce our stress. For example, if we are suffering debilitating anxiety before a court appearance, it might be helpful to examine our belief systems about court. If I am anxious because strange people are going to ask difficult questions, this is a rational belief. If, on the other hand, I am anxious because I am afraid that I will forget my name or my evidence and become a laughing stock – this is an irrational belief.

Increasing your coping resources

An alternative way of reducing stress is to increase our coping resources or get better at looking after ourselves. This will considerably increase our ability to withstand pressure.

Negative versus positive coping mechanisms
Although all coping mechanisms can help reduce stress in the short-term, some will generally have a positive effect: reading, sport, talking with friends, singing and so on. However, others have only a short-term positive effect and, in the long-term, run the risk of leaving us less able to deal with our pressures. These potentially negative coping mechanisms include alcohol, tobacco or other substances; over-eating; over- spending and so on. Although effective in the short-term, if over-indulged they will have a negative rather than a positive effect.

Mixed economy
As we grow older, it is frequently the case that we use fewer and fewer coping methods to look after ourselves. We often find that we come to rely on two or three main ones. This makes us particularly vulnerable to stress if, for some reason, we are deprived of one of these coping mechanisms. So, if my main coping mechanisms involve talking with my partner and walking my dog, what happens when I get divorced and when my pet dies – my normal feelings of grief are exaggerated by having no mechanism left to look after myself. However, if we can develop different coping mechanisms that we can use in all circumstances, we find ourselves in a positive position.

How could you develop and expand your coping resources?

Using staff care/counselling schemes

Several social welfare agencies have now established staff care/counselling schemes for those members of staff who feel severely stressed. One problem for these schemes is that they tend to be under-used or only used when stress leads to severe debilitation. Because everyone has the potential to suffer from stress, everyone also has the potential to need outside investment. This outside investment, particularly if delivered early enough, can be most effective in reducing stress.

Helping others

Recognising stress in others

One of the most effective ways of helping a colleague who is stressed is by recognising it, 'normalising' it, attempting to share part of their pressure and being available for them to talk to.

Making a link without making demands

If a colleague or a team member has actually gone off sick with stress, it is most important to maintain a link with them to try to reduce their isolation. But it is important at the same time to ensure that this contact is not seen as an additional pressure by the other person. Making contact without making demands is the best help that can be offered.

Encouraging appropriate use of the staff care scheme

If a colleague has gone off sick with stress (or is about to do so) it is often helpful to encourage them to consider using an appropriate staff care scheme. The positive benefits of this service and the normality of needing outside investment should be emphasised. All thoughts of stigma, failure or weakness should be challenged.

■ 5. Protecting ourselves from stress: forming a partnership against stress

Although stress is perceived and felt individually, it is not just an individual phenomenon. It is, in essence, a work hazard. Like any other serious work hazard, dealing with stress is a collective problem involving cooperative effort. Stress can be reduced and controlled, its most harmful effects avoided, but this involves cooperative effort from, and a partnership between:

- Your employer
- Your human resources department
- Your union/professional association
- Your manager
- Your team
- Your family
- and (most importantly of all) YOURSELF!

If this partnership is working well, help and support can flow from any part of it. Support is any kind of positive contact, help, assistance or just being there to listen. The questions that this partnership might ask itself include:

- What are the stressors that the staff member/team are suffering at the moment?
- How serious are these stressors?
- How can they be removed or reduced?
- How can we help colleagues where stress is so serious it requires outside assistance?
- How can we reduce the general levels of stress experienced in the workplace on a long-term basis?

Conclusions 5

- Stress is a costly and complex matter which has tremendous destructive potential – for individuals, groups of staff, the organisation and its clients, customers or service users.
- The fact that the negative effects of stress are so widespread underlines the need for us to tackle the problem collectively – to develop a *partnership.*
- We must move beyond the traditional view of stress which sees it simply as a sign of weakness in the individual, and recognise it as a complex interpersonal and organisational phenomenon.
- Blaming the individual simply creates additional pressures and therefore does more harm than good.
- This is not to say that individuals do not have a part to play in stress management – clearly we all do – but there is a danger in overemphasising the individual's role and failing to see the broader picture.
- A narrow perspective on stress is therefore a significant trap to avoid falling into.

In view of this broader perspective, the roles of all concerned need to be reviewed in order to move away from the mistakes arising from traditional oversimplified approaches to this complex area.

The manager's role

The manager's role in meeting the stress challenge can be summarised as:

- Helping to keep pressures within the optimal middle range – not too much, not too little.
- Recognising, respecting and enhancing the coping methods used by staff and encouraging or introducing others as appropriate.
- Providing and facilitating formal and informal support.
- Recognising and addressing barriers to effective stress management.
- Working in a proactive way by creating an open and supportive work atmosphere.
- Recognising one's own stress management needs and taking the necessary steps to meet them – an over-stressed manager will be less well-equipped to support over-stressed staff.

The trainer's role

The role of the trainer can be summarised as:

- Identifying training and staff development needs in relation to stress management and related matters.
- Ensuring that stress and staff care issues feature prominently within training and staff development strategies and plans.
- Recognising, respecting and enhancing the coping methods used by staff and encouraging, facilitating or introducing others as appropriate.
- Ensuring that all training events are positive and constructive, geared towards empowerment and enhanced levels of practice, and therefore not oppressive or involving undue pressures.
- Using one's influence within the organisation to promote an awareness of stress issues and the need for a sustained, systematic approach to staff care.
- Recognising one's own stress management needs and taking the necessary steps to meet them – an over-stressed trainer will be less well-equipped to support over-stressed staff.

The practitioner's role

The role of the practitioner can be summarised as:

- Recognising the range of stressors that you face in order to be better equipped to deal with them.
- Adopting a realistic perspective towards these stressors so that they are neither exaggerated nor swept under the carpet.
- Taking stock of the coping methods and resources available to be drawn upon so that, where necessary, these can be developed, consolidated and supplemented.
- Identifying potentially problematic coping methods (drinking to excess, for example) and guarding against them.
- Identifying, and being prepared to draw upon, the range of potential supports available.
- Offering support to colleagues where and when necessary, and generally contributing to an atmosphere and ethos of support and staff care.

Tackling stress is neither easy nor straightforward but, as we have seen, it is nonetheless vitally important that the challenge is taken seriously by all concerned – managers, trainers and practitioners – and given the investment of time and energy it needs.

Two comforts, though, that can be taken from this difficult challenge are:

1. Over time, it gets easier as knowledge, skills and confidence are built up.
2. As staff respond positively to your efforts to deal with stress, there is tremendous job satisfaction to be gained from seeing what a difference a positive approach to the stress challenge can make.

As new changes occur and new demands and challenges emerge, the battle against stress has to be fought anew, thereby underlining the need to have a clear strategy for dealing with stress and a shared commitment to tackling problems as and when they arise. Stress has a nasty habit of dividing and conquering, and so the need to work together in partnership is one that cannot be emphasised enough. Good luck in developing and strengthening that partnership in meeting the stress challenge together.

References

Agha, A.O., Mueller, C.W. and Price, J.L. (1993) 'Determinants of Employee Job Satisfaction: An Empirical Test of a Causal Model'. *Human Relations,* 46 (8), 1007-1027.

Arroba, T. and James, K. (1987) *Pressure at Work: A Survival Guide,* London, McGraw-Hill.

Ashforth, B.E. and Humphrey, R.H. (1993) 'Emotional Labor in Service Roles: The Influence of Identity', *Academy of Management Review* 18 (1).

Bates, J. and Pugh, R.G. (1995) 'A Case Study which Demonstrates the Innovative Use of IT with People with Learning Disabilities', *New Technology in the Human Services,* 8 (1).

Bone, C. (1994) *The Modern Quality Management Manual,* Harlow, Longman.

Cartwright, S. and Cooper, C. (1994) *No Hassle! Taking the Stress out of Work,* London, Century Business Books, Random House.

Cooper, C.L. (1995a) 'An Assessment of Stress Counselling in the Workplace', UMIST Research Focus, 11 December.

Cooper, C.L. (1995b) Paper presented at the Institute of Personnel and Development conference, Harrogate.

Cooper, C.L. and Williams, S. (eds) (1994) *Creating Healthy Work Organizations,* Chichester, Wiley.

Fingret, A. (1994) 'Developing a Company Mental Health Plan', in Cooper and Williams (1994).

Hochschild, A. (1983) *The Managed Heart: The Commercialization of Human Feeling,* New York, Anchor.

Jones, F., Fletcher, B.C. and Ibbetson, K. (1991) 'Stressors and Strains among Social Workers: Demands, Supports, Constraints and Psychological Health', *British Journal of Social Work,* 21.

Keita, G.P. and Sauter, S.L. (eds) (1992) *Work and Well-Being: An Agenda for the 1990s,* Washington, DC, American Psychological Association.

Harding, H. (1987) *Management Appreciation,* London, Pitman.

Maslach, C. and Jackson, S. (1981) *The Maslach Burnout Inventory,* Palo Alto, Cal., Consulting Psychology Press.

NCH (1993) *NCH Factfile: Children in Britain 1992,* London, NCH.

Patel, C. (1989) *The Complete Guide to Stress Management,* London, Macdonald.

Pottage, D. and Evans, M. (1992) *Workbased Stress: Prescription is not the Cure,* London, NISW.

Quick, J.C. and Quick, J.D. (1984) *Organisational Stress and Preventative Management,* New York, McGraw Hill.

Murphy, M. (1995) *Working Together in Child Protection,* Aldershot, Arena.

Reddy, M. (1992) *Counselling: Its Value to Business,* London, HMSO.

Roskies, E. (1983) 'Stress Management: Averting the Evil Eye', *Contemporary Psychology* 28 (7).

Ross, R.R. and Altmaier, E.M. (1994) *Intervention in Occupational Stress,* London, Sage.

Scott, M.J. and Stradling, S.G. (1992) *Counselling for Post-Traumatic Stress Disorder,* London, Sage.

Thompson, N. (1991) 'Breaking Cycles', *Community Care,* 31 January.

Thompson, N. (1992) *Existentialism and Social Work,* Aldershot, Avebury.

Thompson, N. (1993) *Anti-Discriminatory Practice,* London, Macmillan.

Thompson, N. (1996) *People Skills,* London, Macmillan.

Thompson, N. and Bates, J. (1995) 'In-service Training: Myth and Reality', *Curriculum* 16 (1).

Thompson, N., Murphy, M. and Stradling, S. (1994) *Dealing with Stress,* London, Macmillan.

Thompson, N., Stradling, S., Murphy, M. and O'Neill, P. (1996) 'Stress and Organizational Culture', *British Journal of Social Work* 26 (2).

Wagenaar, J. and La Forge, J. (1994) 'Stress Counselling Theory and Practice: A Cautionary Review', *Journal of Counselling and Development,* September/October.

Weinstein, N. (1988) 'The precaution adoption process', *Health Psychology,* 7, 55-386.

Wharton, A.S. (1993) 'The Affective Consequences of Service Work: Managing Emotions on the Job', *Work and Occupations* 20 (2).

Finding out more

Argyle, M. (1989) *The Social Psychology of Work,* 2nd edn, Harmondsworth, Penguin.

Arroba, T. and James, K. (1987) *Pressure at Work: A Survival Guide,* London, McGraw-Hill.

Burnard, P. (1991) *Coping with Stress in the Health Professions,* London, Chapman and Hall.

Cartwright, S. and Cooper, C. (1994) *No Hassle! Taking the Stress out of Work,* London, Century Business Books, Random House.

Cooper, C.L., Cooper, R.D. and Baker, L. (1988) *Living with Stress,* Penguin, Harmondsworth.

Cooper, C.L. and Williams, S. (eds) (1994) *Creating Healthy Work Organizations,* Chichester, Wiley.

Cox, T. (1978) *Stress,* London, Macmillan.

Curtis, C. and Metcalf, J. (1992) *Becoming a Care Supervisor,* Edinburgh, Churchill Livingstone.

Education Service Advisory Committee (1990) *Managing Occupational Stress: A Guide for Managers and Teachers in the Schools Sector,* London, HMSO.

Hawkins, P. and Shohet, R. (1991) *Supervision in the Helping Professions,* Buckingham, Open University Press.

Hope, P. and Pickles, T. (1995) *Performance Appraisal: A Handbook for Managers in Public and Voluntary Organisations,* Lyme Regis, Russell House Publishing

Jee, M. and Reason, L. (1988) *Action on Stress at Work,* London, Health Education Authority.

Keita, G.P. and Sauter, S.L. (eds) (1992) *Work and Well-Being: An Agenda for the 1990s,* Washington, DC, American Psychological Association.

Leadbetter, D. and Trewartha, R. (1996) *Handling Aggression and Violence at Work,* Lyme Regis, Russell House Publishing.

Looker, T. and Gregson, O. (1989) *Stresswise,* Sevenoaks, Hodder and Stoughton.

Patel, C. (1989) *The Complete Guide to Stress Management,* London, Macdonald.

Pottage, D. and Evans, M. (1992) *Workbased Stress: Prescription is not the Cure,* London, NISW.

Powell, T.J. and Enright, S.J. (1990) *Anxiety and Stress Management,* London, Routledge.

Rees, S. and Graham, R.S. (1991) *Assertion Training,* London, Routledge.

Ross, R.R. and Altmaier, E.M. (1994) *Intervention in Occupational Stress,* London, Sage.

Scott, M.J. and Stradling, S.J. (1992) *Counselling for Post-traumatic Stress Disorder,* London, Sage.

Selye, H. (1974) *Stress without Distress,* London, Corgi.

Thompson, N. (1991) 'Breaking Cycles', *Community Care* 31 January.

Thompson, N., Murphy, M. and Stradling, S. (1994) *Dealing with Stress,* London, Macmillan.

Thompson, N., Stradling, S., Murphy, M. and O'Neill, P. (1996) 'Stress and Organizational Culture', *British Journal of Social Work* 26 (5).

Totman, R. (1990) *Mind, Stress and Health,* London, Souvenir Press.

Appendix: OHP templates

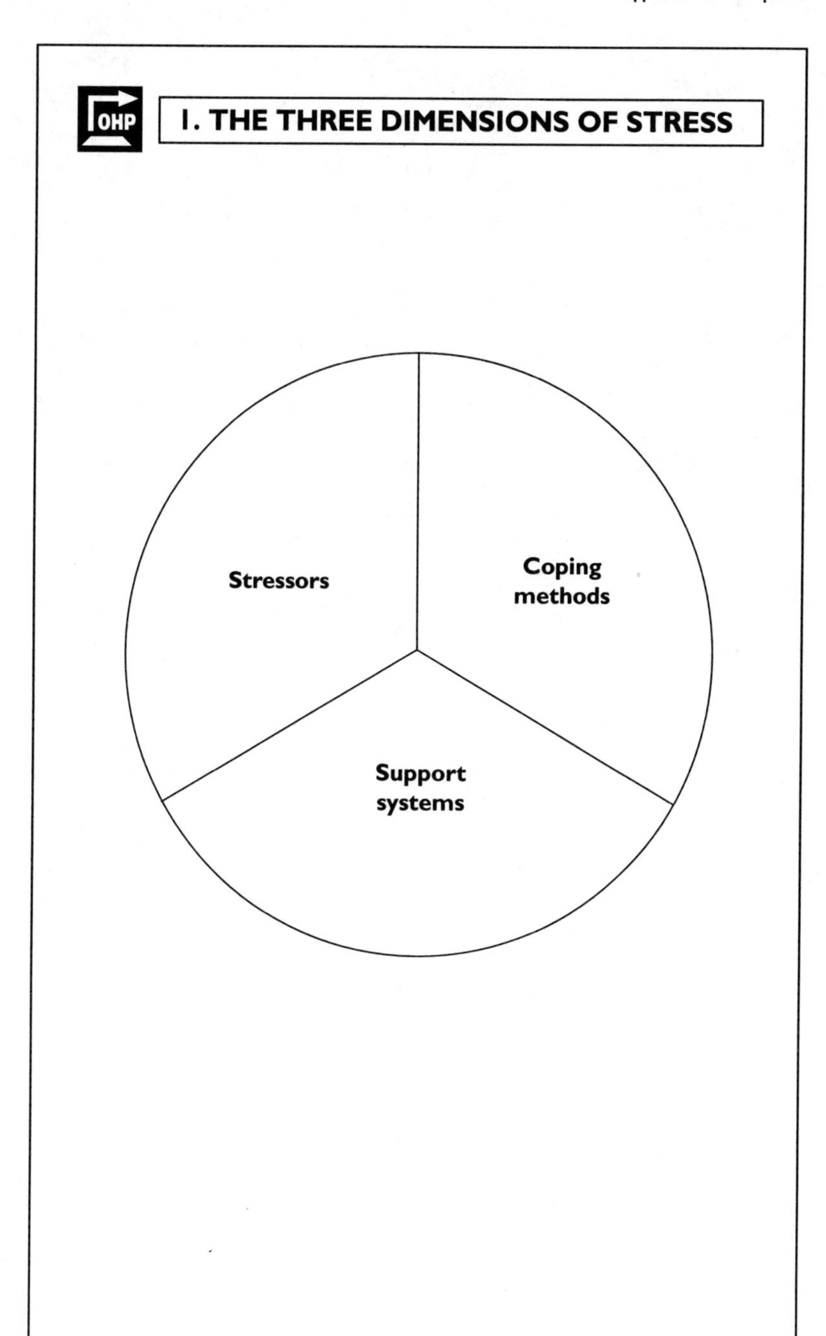
OHP
1. THE THREE DIMENSIONS OF STRESS
Stressors
Coping methods
Support systems

OHP

2. BUILDING PARTNERSHIPS

What can I do?	What can they do?

3. BOUNDARIES OF RESPONSIBILITY

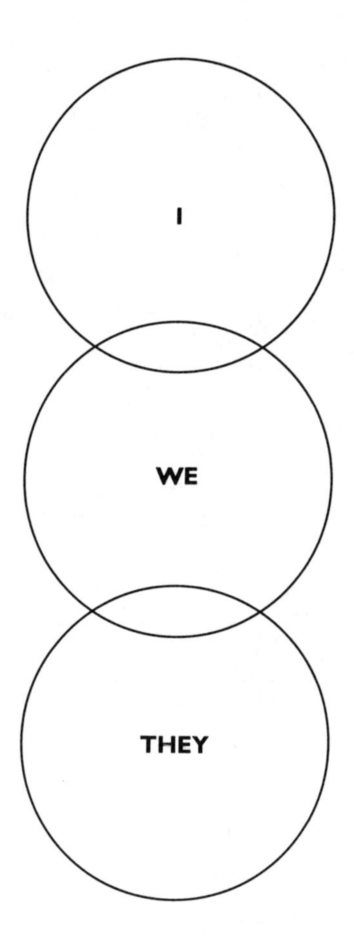

Based on training materials prepared by Tony Morrison.

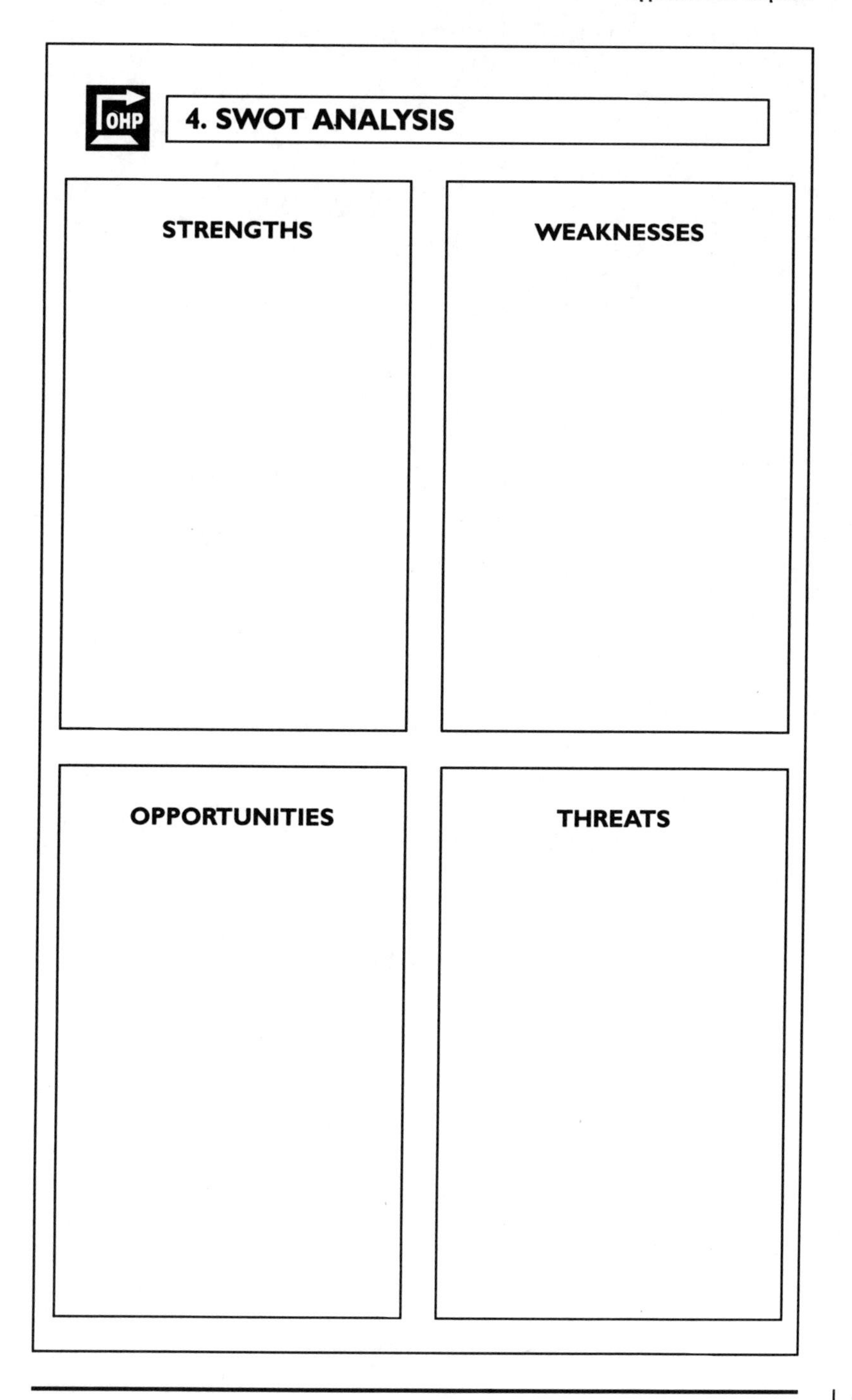
OHP
4. SWOT ANALYSIS
STRENGTHS
WEAKNESSES
OPPORTUNITIES
THREATS

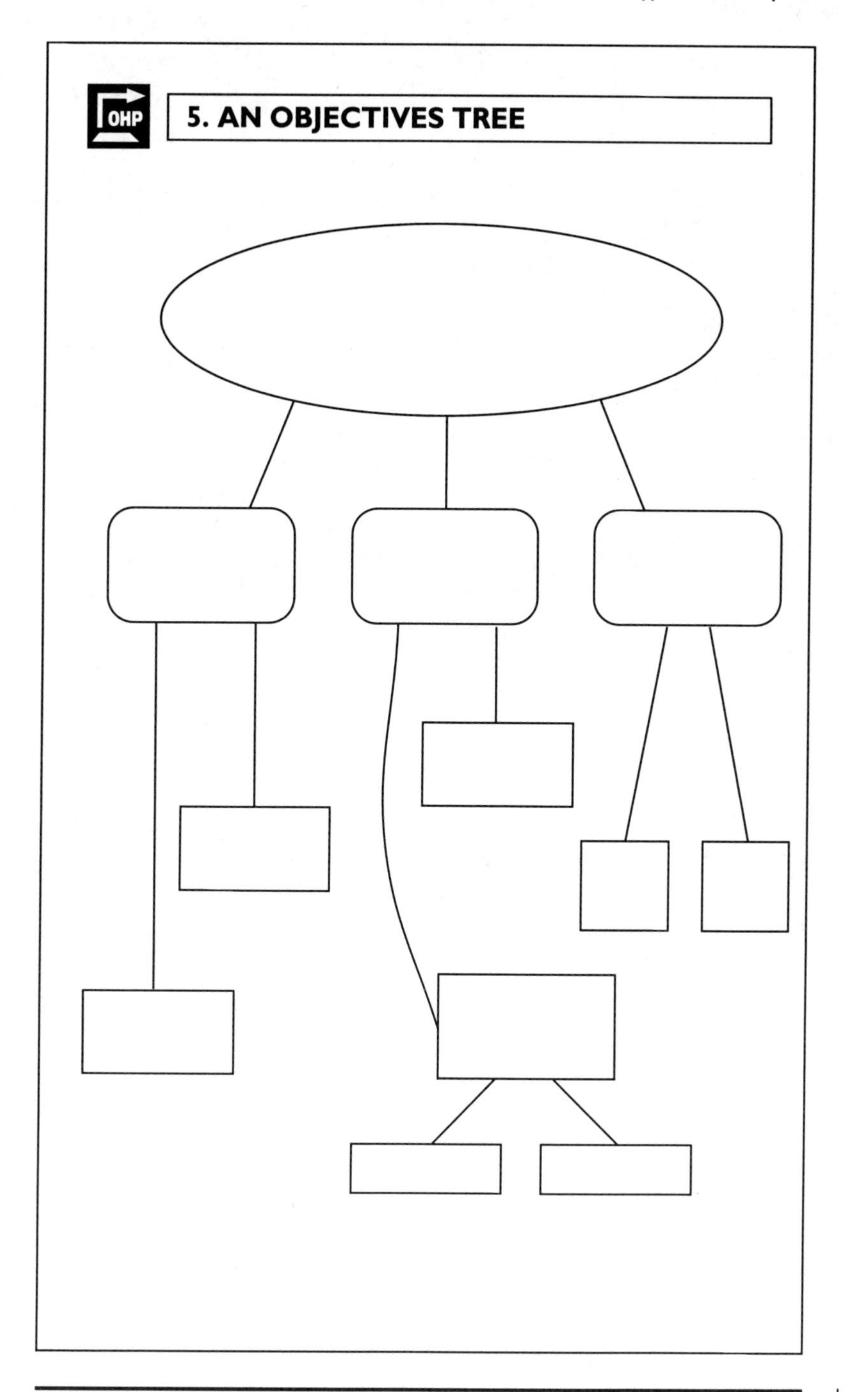
OHP
5. AN OBJECTIVES TREE

6. WHAT, HOW, WHEN?

1. **What are you trying to achieve?**
2. **How are you going to achieve it?**
3. **How will you know when you have achieved it?**

Russell House Publishing resources

We hope that trainers and managers in social welfare and informal education will find these resources helpful in their work.

- ☐ Anti-racist Work with Young People: European Experiences and Approaches
- ☐ Black Professionals in Welfare
- ☐ Chasing Rainbows: Children, Divorce and Loss
- ☐ Community Care Assessment Casebook
- ☐ Creative Outdoor Work with Young People
- ☐ Employment Practice and Procedures in Youth and Community Work
- ☐ The Foster Carer's Handbook
- ☐ Handling Aggression and Violence at Work
- ☐ New Youth Games Book
- ☐ Performance Appraisal: A Handbook for Managers in Public and Voluntary Organisations
- ☐ Promoting Partnerships through Consultation
- ☐ Quicksilver: Adventure Games, Initiative Problems, Trust Activities and a Guide to Effective Leadership
- ☐ Sexuality, Young People and Care: Creating Positive Contexts for Training, Policy and Development
- ☐ What Works in Family Mediation
- ☐ World Youth Games
- ☐ Working in Partnership: The Probation Service and the Voluntary Sector
- ☐ Working with Men

To obtain more information about these and other RHP titles, photocopy this page, tick the relevant boxes and mail to:

Russell House Publishing Ltd.,
38 Silver Street
LYME REGIS
Dorset DT7 3HS

Or telephone us on (01297) 443948.